by James Richard

ALGEBRA

WORKBOOK 2

January 2020

Contents

roots

Definition

: $\sqrt[n]{a} = x \Rightarrow x^n = a$

$\sqrt[n]{a^n} = a$, (n is an odd number)

$\sqrt[n]{a^n} = a$, (n is an even number)

$a < 0 \Rightarrow \sqrt[n]{a^n} \in R$ (n is an even number)

(Example):

$$\sqrt{0.25} + \sqrt{1.21} + \sqrt{1.44} = ?$$

(Solution):

$$\sqrt{0.5^2} + \sqrt{1.1^2} + \sqrt{1.2^2}$$

=0.5+1.1+1.2=2.8

(Example):

$$\frac{\sqrt[3]{(-3)^3} - \sqrt[6]{(-3)^6}}{-\sqrt{(-2)^2} - \sqrt[6]{-24}} = ?$$

A)-6   B)-3   C)3   D)6

(Solution):

$$\frac{\sqrt[3]{(-3)^3}-\sqrt[6]{(-3)^6}}{-\sqrt{(-2)^2}-\sqrt[6]{(-3)^5}}$$

$$=\frac{-3-3}{-2+3}=\frac{-6}{1}=-6$$

(Example):

$$\frac{\sqrt{2,7}+\sqrt{5,4}}{\sqrt{0,-1}+\sqrt{0,4}}=?$$

(Solution):

$$\frac{\sqrt{2+\frac{7}{9}}+\sqrt{5+\frac{4}{9}}}{\sqrt{\frac{1}{9}}+\sqrt{\frac{4}{9}}}$$

$$=\frac{\sqrt{2+\frac{7}{9}}+\sqrt{5+\frac{4}{9}}}{\frac{1}{3}+\frac{2}{3}}=\frac{\frac{5}{3}+\frac{7}{3}}{\frac{3}{3}}=\frac{12}{3}=4$$

PROPERTIES

1. $\sqrt[m]{a^n} = a^{\frac{n}{m}}$

(Example):

$$\sqrt[3]{3^{x+6}} + 9.\sqrt[3]{8.3^x} = 9 \implies x =?$$

A)-3   B)-1   C)1   D)3   E)4

(Solution):

$$\sqrt[3]{3^x.3^6} + 9.\sqrt[3]{8.3^x} = 9$$

$$3^2\sqrt[3]{3^x}. + 2.9.\sqrt[3]{3^x} = 9$$

$$27\sqrt[3]{3^x} = 9$$

$$3^{\frac{x}{3}} = 3^{-1}$$

$$\frac{x}{3} = -1 \implies x = -3$$

-Answer A

2. $\sqrt[m]{\sqrt[n]{a}} = \sqrt[m.n]{a}$

$$\sqrt[-x-3]{\sqrt[x-3]{81}} = \sqrt[4]{3} \implies x =?$$

A)1   B)3   C)5   D)7   E)9

(Solution):

$$\sqrt[x^2-9]{81} = \sqrt[4]{3}$$
$$3^{\frac{4}{x^2-9}} = 3^{\frac{1}{4}}$$
$$\frac{4}{x^2-9} = \frac{1}{4}$$
$$x^2 - 9 = 16 \implies x^2 = 25$$
$$X = 5$$

3. $\sqrt{a + \sqrt{b}} = \sqrt{\dfrac{a+\sqrt{a^2-b}}{2}} + \sqrt{\dfrac{a-\sqrt{a^2-b}}{2}}$

4. $\sqrt{a - \sqrt{b}} = \sqrt{\dfrac{a+\sqrt{a^2-b}}{2}} - \sqrt{\dfrac{a-\sqrt{a^2-b}}{2}}$

(Example):

$$\sqrt{3 - 2\sqrt{2}} = ?$$

(Solution):

4

$$\sqrt{3 - 2\sqrt{2}} = \sqrt{(\sqrt{2} - \sqrt{1})^2}$$
$$= \sqrt{2} - 1$$

(Example):

$$(2 - \sqrt{3}).\sqrt{7 + 4\sqrt{3}} = ?$$

A)1   B)2   C)3   D)4   E)5

(Solution):

$$= (2 - \sqrt{3}).\sqrt{7 + 2.2\sqrt{3}}$$
$$= (2 - \sqrt{3}).\sqrt{7 + 2\sqrt{12}}$$
$$= (2 - \sqrt{3}).(\sqrt{4} + \sqrt{3})$$
$$= (2)^2 - (\sqrt{3})^2$$
$$= 4\text{-}3 = 1$$

Answer A

(Example):

$$\sqrt{14 + 8\sqrt{3}} = 2\sqrt{2} + b \Rightarrow 2\sqrt{3}.b = ?$$

A) $\sqrt{2}$ B)$2\sqrt{2}$ C)$3\sqrt{2}$ D)$4\sqrt{2}$ E)$6\sqrt{2}$

(Solution):

$$\sqrt{14 + 2.4\sqrt{3}} = 2\sqrt{2} + b$$
$$\sqrt{14 + 2\sqrt{48}} = 2\sqrt{2} + b$$
$$\sqrt{8} + \sqrt{6} = \sqrt{8} + b$$
$$b = \sqrt{6}\, dir$$
$$2\sqrt{3} \cdot b = 2\sqrt{3}\sqrt{6}$$
$$= 2\sqrt{18} = 2.3\sqrt{2} = 6\sqrt{2}$$

Answer E

(Example):

$$\sqrt{5} = x \Rightarrow \sqrt{9 - 4\sqrt{5}} = ?$$

A)x-1   B)x-2   C)x+1

D)x+2   E)x+4

(Solution):

$$\sqrt{9 - 2.2\sqrt{5}} = \sqrt{9 - 2\sqrt{20}}$$

$$= \sqrt{5} - \sqrt{4}$$

$$= x - 2$$

Answer B

(Example):

$$\sqrt{2 + \sqrt{3}} - \sqrt{4 - 2\sqrt{3}}$$

A)1   B)2   C)4   D)6   E)8

(Solution):

$$\sqrt{2 + \sqrt{3}} - \sqrt{4 - 2\sqrt{3}} = \sqrt{3} + 1 - (\sqrt{3} - 1)$$

$$= \sqrt{3} + 1 - \sqrt{3} + 1$$

$$= 2$$

Answer B

(Example):

$$\sqrt{2 + \sqrt{3}} - \sqrt{2 - \sqrt{3}}$$

A) $\sqrt{2}$ B) $2\sqrt{2}$ C) 4   D) $4\sqrt{2}$ E) 8

(Solution):

$$\sqrt{2+\sqrt{3}} - \sqrt{2-\sqrt{3}}$$

$$= \left(\sqrt{\frac{3}{2}} + \sqrt{\frac{1}{2}}\right) - \left(\sqrt{\frac{3}{2}} - \sqrt{\frac{1}{2}}\right)$$

$$= \sqrt{\frac{3}{2}} + \sqrt{\frac{1}{2}} - \sqrt{\frac{3}{2}} + \sqrt{\frac{1}{2}}$$

$$= 2\sqrt{\frac{1}{2}} = \sqrt{4 \cdot \frac{1}{2}} = \sqrt{2}$$

-Answer A

5. $x = \sqrt[n]{a\sqrt[n]{a\sqrt[n]{a \dots \dots \dots}}} \Rightarrow x = \sqrt[n-1]{a}$

6. $a. \sqrt[n]{x} + b. \sqrt[n]{x} - c. \sqrt[n]{x} = (a+b-c). \sqrt[n]{x}$

(Example):

$$\sqrt{40} + \sqrt{\frac{2}{5}} - \sqrt{\frac{2}{5}} = \frac{a\sqrt{40}}{20} \Rightarrow a = ?$$

A) 9   B) 11   C) 13   D) 15   E) 17

8

(Solution):

$$\sqrt{40 \cdot \frac{2}{5}} + \sqrt{\frac{2}{5} \cdot \frac{2}{5}} - \sqrt{\frac{5}{2} \cdot \frac{2}{5}} = \frac{a\sqrt{40 \cdot \frac{2}{5}}}{20}$$

$$\sqrt{16} + \frac{2}{5} - \sqrt{1} = \frac{40.a}{20}$$

$$3 + \frac{2}{5} = \frac{a}{5}$$

$$a = 17$$

-Answer E

7. $\sqrt[n]{a} \cdot \sqrt[n]{b} = \sqrt[n]{a \cdot b}$

(Example):

$$\sqrt{2}\sqrt{3} \ldots \ldots \ldots \sqrt{n} = 2\sqrt{30} \Rightarrow n = ?$$

A)2   B)3   C)5   D)8   E)16

(Solution):

$$\sqrt{2}\sqrt{3} \ldots \ldots \ldots \sqrt{n} = \sqrt{4.30}$$

$$\sqrt{n!} = \sqrt{120}$$

$$l20 = 5l$$

$$n! = 5!$$

$$\Rightarrow n = 5$$

-Answer C

8. $\dfrac{\sqrt[n]{a}}{\sqrt[n]{b}} = \sqrt[n]{\dfrac{a}{b}}$

9. $\dfrac{a}{\sqrt{b}} = \dfrac{a.\sqrt{b}}{\sqrt{b}.\sqrt{b}} = \dfrac{a\sqrt{b}}{b}$

10. $\dfrac{a}{\sqrt{b}+\{\sqrt{c}} = \dfrac{a.(\sqrt{b}-\sqrt{c})}{(\sqrt{b}-\sqrt{c}).(\sqrt{b}+\sqrt{c})} = \dfrac{a.(\sqrt{b}-\sqrt{c})}{b-c}$

11. $\dfrac{a}{\sqrt{b}-\sqrt{c}} = \dfrac{a.(\sqrt{b}+\sqrt{c})}{(b-\sqrt{c}).(b+\sqrt{c})} = \dfrac{a.(b+\sqrt{c})}{b^2-c}$

(Example):

$$\dfrac{1}{\sqrt{3}-\sqrt{2}} - \dfrac{2}{\sqrt{2}} = ?$$

A)$\dfrac{\sqrt{3}}{2}$ B) $\sqrt{3}$ C)$2\sqrt{3}$ D) $3\sqrt{3}$ E) $4\sqrt{3}$

(Solution):

$$\dfrac{1}{\sqrt{3}-\sqrt{2}} - \dfrac{2}{\sqrt{2}}$$

$$= \dfrac{\sqrt{3}+\sqrt{2}}{3-2} - \dfrac{2\sqrt{2}}{2}$$

$$= \sqrt{3} + \sqrt{2} - \sqrt{2} = \sqrt{3}$$

-Answer B

(Example):

$$\frac{4}{\sqrt{5}-1} + \frac{1}{\sqrt{2}-1} - \frac{3}{\sqrt{5}-\sqrt{2}} = ?$$

A)1   B)2   C)3   D)4   E)5

(Solution):

$$\frac{4}{\sqrt{5}-1} + \frac{1}{\sqrt{2}-1} - \frac{3}{\sqrt{5}-\sqrt{2}}$$

$$= \frac{4(\sqrt{5}+1)}{5-1} + \frac{\sqrt{2}+1}{2-1} - \frac{3(\sqrt{5}+\sqrt{2})}{5-2}$$

$$= \sqrt{5} + 1 + \sqrt{2} + 1 - \sqrt{5} - \sqrt{2} = 2$$

-Answer B

TEST WITH SOLUTION

(Example):

1. $3\sqrt{147} + 2\sqrt{75} - 5\sqrt{108} = ?$

A)0   B) $\sqrt{3}$ C)-2$\sqrt{7}$ D) 2$\sqrt{7}$ E)6$\sqrt{3}$

Çözüm (Solution):

$$3\sqrt{147} + 2\sqrt{75} - 5\sqrt{108}$$
$$= 3\sqrt{49.3} + 2\sqrt{25.3} - 5\sqrt{36.3}$$
$$= 3.7\sqrt{3} + 2.5\sqrt{3} - 5.6\sqrt{3}$$
$$= 21\sqrt{3} + 10\sqrt{3} - 30\sqrt{3}$$
$$= \sqrt{3}$$

-Answer B

2. $3\sqrt{40} - \sqrt{250} + \frac{20}{\sqrt{10}} = ?$

A) $\sqrt{10}$ B) $3\sqrt{10}$ C) $5\sqrt{10}$ D) $2\sqrt{20}$ E) $\sqrt{50}$

(Solution):

$$3\sqrt{40} - \sqrt{250} + \frac{20}{\sqrt{10}}$$

$$= 3\sqrt{4.10} - \sqrt{25.10} + \frac{20.\sqrt{10}}{10}$$

$$= 3.2\sqrt{10} - 5.\sqrt{10} + 2.\sqrt{10}$$

$$= 6\sqrt{10} - 5\sqrt{10} + 2.\sqrt{10}$$

$$= (6 - 5 + 2)\sqrt{10} = 3\sqrt{10}$$

-Answer B

3. $18\sqrt{\frac{8}{27}} - \sqrt{150} = ?$

12

A)0   B)$-\sqrt{6}$ C)$\sqrt{6}$ D) 2-$\sqrt{3}$ E)2$\sqrt{3}$

(Solution):

$$18\sqrt{\frac{8}{27}} - \sqrt{150} = 18\sqrt{\frac{4.2}{9.3}} - \sqrt{25.3}$$

$$= 18.\frac{2\sqrt{2}}{3\sqrt{3}} - 5\sqrt{6}$$

$$= 12.\frac{\sqrt{2}}{\sqrt{3}} - 5\sqrt{6}$$

$$= 12.\frac{\sqrt{6}}{3} - 5\sqrt{6} = -\sqrt{6}$$

-Answer B

4. $\frac{1}{2}\sqrt{32} - \frac{1}{3}\sqrt{18} + \frac{\sqrt{6}}{\sqrt{3}} = ?$

A)0   B)-1 C)$\sqrt{2}$ D) 2$\sqrt{2}$ E)$\sqrt{3}$

(Solution):

$$\frac{1}{2}\sqrt{32} - \frac{1}{3}\sqrt{18} + \frac{\sqrt{6}}{\sqrt{3}} = \frac{1}{2}\sqrt{16.2} - \frac{1}{3}\sqrt{9.2} + \frac{\sqrt{18}}{3}$$

$$= \frac{1}{2}.4\sqrt{2} - \frac{1}{3}.3\sqrt{2} + \frac{3\sqrt{2}}{3}$$

$$= 2\sqrt{2} - \sqrt{2} + \sqrt{2}$$

-Answer D

5. $\sqrt{\dfrac{1}{16}+\dfrac{1}{9}}\cdot\sqrt{\dfrac{1}{9}-\dfrac{1}{25}}=?$

A)1   B)$\dfrac{1}{3}$ C)16   D)$\dfrac{1}{9}$ E)$\dfrac{1}{20}$

$$\sqrt{\frac{1}{16}+\frac{1}{9}}\cdot\sqrt{\frac{1}{9}-\frac{1}{25}}=\sqrt{\frac{25}{144}}\cdot\sqrt{\frac{16}{225}}$$

$$=\frac{5}{12}\cdot\frac{4}{15}=\frac{20}{180}=\frac{1}{9}$$

-Answer D

11. 4. $\sqrt{\dfrac{3}{2}}-\sqrt{54}+3\sqrt{\dfrac{2}{3}}=?$

A)6   B)$\sqrt{6}$ C)$-\sqrt{6}$ D)0   E)$6\sqrt{6}$

(Solution):

4. $\sqrt{\dfrac{3}{2}}-\sqrt{54}+3\sqrt{\dfrac{2}{3}}$

$=\dfrac{4\cdot\sqrt{6}}{2}-\sqrt{\mathbf{9.6}}+\dfrac{3\cdot\sqrt{6}}{3}$

$=2\cdot\sqrt{6}-3\cdot\sqrt{6}+\sqrt{6}=0$

14

12. $\sqrt{1.44} - \sqrt{19.6} + \sqrt{\dfrac{490}{25}} = ?$

$A)0.2$ B)0.9   C)1   D)1.2   E)1.4

(Solution):

$\sqrt{\dfrac{144}{100}} - \sqrt{\dfrac{196}{10}} + \sqrt{\dfrac{490}{25}} = \dfrac{12}{10} - \dfrac{14}{\sqrt{10}} + \dfrac{7\sqrt{10}}{5}$

$= \dfrac{12}{10} - \dfrac{14.\sqrt{10}}{10} + \dfrac{7\sqrt{10}}{5}$

$= \dfrac{12}{10} - \dfrac{7\sqrt{10}}{5} + \dfrac{7\sqrt{10}}{5} = \dfrac{12}{10} = \mathbf{1.2}$

-Answer D

13. $\left(1 - \dfrac{1}{\sqrt{2}}\right)\left(\dfrac{1}{2-\sqrt{2}}\right) = ?$

(Solution):

$\left(1 - \dfrac{1}{\sqrt{2}}\right) \cdot \left(\dfrac{1}{2-\sqrt{2}}\right) = \left(1 - \dfrac{\sqrt{2}}{2}\right)\left(\dfrac{1}{2-\sqrt{2}}\right)$

$= \dfrac{2-\sqrt{2}}{2} \cdot \dfrac{1}{2-\sqrt{2}}$

$$= \frac{1}{2}$$

-Answer A

14. $\dfrac{2}{1-\dfrac{\frac{\sqrt{2}}{1}}{1-\frac{1}{\sqrt{2}-1}}}$ =?

A)-2   **B)-1**   **C)0**   **D)1**   **E)2**

Solution

$$\frac{2}{1-\dfrac{\frac{\sqrt{2}}{1}}{1-\frac{1}{\sqrt{2}-1}}} = \frac{2}{1-\dfrac{\frac{\sqrt{2}}{1}}{1-\frac{\sqrt{2}+1}{1}}}$$

$(\sqrt{2}+1)$

$1-\frac{\sqrt{2}}{1}$ -

$1-\dfrac{1}{\sqrt{2}-1}$ $1-\dfrac{1}{\sqrt{2}-1}$

$(\sqrt{2}+1)$

$$=\frac{2}{1-\dfrac{\frac{\sqrt{2}}{1}}{1-\sqrt{2}-1}}$$

$$=\frac{2}{1-\dfrac{\sqrt{2}}{-\sqrt{2}}}$$

$$=\frac{2}{1+1}=1$$

-Answer

15. $\dfrac{\sqrt{45+20}}{4\sqrt{20}-\sqrt{5}}$

A)$\dfrac{5}{4}$ B) $\dfrac{5}{3}$ C)$\dfrac{5}{2}$ D) $\dfrac{5}{6}$ E)$\dfrac{5}{7}$

(Solution):

$$\dfrac{\sqrt{45}+\sqrt{20}}{4\sqrt{20}-\sqrt{5}} = \dfrac{\sqrt{9.5}+\sqrt{4.5}}{4\sqrt{4.5}-\sqrt{5}} \qquad =\dfrac{3.\sqrt{5}+2.\sqrt{5}}{8.\sqrt{5}-\sqrt{5}} = \dfrac{5.\sqrt{5}}{7\sqrt{5}-}=\dfrac{5}{7}$$

16. $\dfrac{\sqrt{3}+\sqrt{2}}{\sqrt{3}-\sqrt{2}} - 2\sqrt{6}$=?

A)1   B)2   C)$\sqrt{6}$ D)4   E)5

(Solution):

$$\dfrac{\sqrt{3}+\sqrt{2}}{\sqrt{3}-\sqrt{2}} - 2\sqrt{6}= \dfrac{(\sqrt{3}+\sqrt{2})\,(\sqrt{3}+\sqrt{2}\,)}{(\sqrt{3}-\sqrt{2}\,)(\sqrt{3}-\sqrt{2}\,)} - 2\sqrt{6}$$

$$=\dfrac{3+\sqrt{66}+\sqrt{6}+2}{1} - 2\sqrt{6}$$

$$=5+2\sqrt{6} - 2\sqrt{6} =5$$

-Answer E

17. $\sqrt{2} + \sqrt{3} - \dfrac{1}{\sqrt{2}+\sqrt{3}}$

A)-$2\sqrt{2}$ B)-$\sqrt{2}$ C) $2\sqrt{2}$ D)$3\sqrt{2}$ E)$4\sqrt{2}$

(Solution):

$\sqrt{2} + \sqrt{3} - \dfrac{1}{\sqrt{2}+\sqrt{3}}$

$(\sqrt{2} - \sqrt{3})$

$=\sqrt{2} + \sqrt{3} - \dfrac{\sqrt{2} - \sqrt{3})}{(\sqrt{2}+\sqrt{3})(\sqrt{2} - \sqrt{3})}$

$=\sqrt{2} + \sqrt{3} - \dfrac{(\sqrt{2} - \sqrt{3})}{-1}$

$=\sqrt{2} + \sqrt{3} + \sqrt{2} - \sqrt{3} = 2\sqrt{2}$

                    -Answer C

18. $\dfrac{10}{\sqrt[3]{25}}$

A)$\sqrt{5}$ B)$\sqrt[3]{5}$ C)$2\sqrt[3]{5}$ D)-)$\sqrt[3]{5}$ E)$5\sqrt{5}$

(Solution):

$\dfrac{10}{\sqrt[3]{25}} = \dfrac{10}{\sqrt[3]{5^2}} = \dfrac{10 \cdot \sqrt[3]{5}}{5} = 2^3\sqrt{5}$

$(\sqrt[3]{5})$ -Answer C

19. $\frac{1}{\sqrt{3}+1} - \frac{3}{\sqrt{3}-1} + \frac{3}{\sqrt{3}} = ?$

A) $-2\sqrt{3}$ **B)** -2 **C)** 1 **D)** 2
E) $2\sqrt{3}$

(Solution):

$\frac{1}{\sqrt{3}+1} - \frac{3}{\sqrt{3}-1} + \frac{3}{\sqrt{3}}$

$(\sqrt{3}-1) \quad \sqrt{3}+1) \quad (\sqrt{3})$

$= \frac{\sqrt{3}-1)}{3-1} - \frac{3.\sqrt{3}+1}{3-1} + \frac{3\sqrt{3}}{3}$

$= \frac{\sqrt{3}-1)}{2} - \frac{3.\sqrt{3}+1}{2} + \sqrt{3})$

$= \frac{\sqrt{3}-1-3\sqrt{3}-3}{2} + \sqrt{3}$

$= \frac{-2\sqrt{3}-4}{2} + \sqrt{3}$

$= \frac{2(-\sqrt{3}-2)}{2} + \sqrt{3}$

$= -\sqrt{3}-2+\sqrt{3}$

$= -2$

-Answer B

19

20. $\dfrac{\left(\sqrt[3]{4-\sqrt[3]{2}}\right)}{\sqrt[3]{2}-1}=?$

A)2   B)4   C)6   D)8   E)10

(Solution):

$$\frac{\left(\sqrt[3]{4-\sqrt[3]{2}}\right)}{\sqrt[3]{2}-1}=\frac{\sqrt[3]{4^2}-\sqrt{2^2}}{\sqrt[3]{2}-1}$$

$$=\frac{\sqrt[3]{16}-2}{\sqrt[3]{2}-1}=\frac{\sqrt[3]{2^3.2}-2}{\sqrt[3]{2}-1}$$

$$=\frac{2\sqrt[3]{2}-2}{\sqrt[3]{2}-1}$$

$$=\frac{2(\sqrt[3]{2}-1)}{\sqrt[3]{2}-1}=2 \qquad\qquad \text{-Answer B}$$

21. $\dfrac{\sqrt{\frac{0.4}{10}}}{\sqrt{0.04}-\sqrt{0.16}}=?$

A)4   B)2   C)0.5   D)-1   E)-1.5

(Solution):

$$\frac{\sqrt{\frac{0.4}{10}}}{\sqrt{0.04}-\sqrt{0.16}} = \frac{\sqrt{0.04}}{\sqrt{0.04}-\sqrt{0.16}}$$

$$=\frac{0.2}{0.2-0.4} = \frac{0.2}{-0.2}=\text{-1}$$

-Answer D

22. $\frac{\sqrt{0.64}-\sqrt{1.96}}{\sqrt{0.36}} + 1=?$

A)-2　　　B)-1　　　C)0　　　D)1　　　E)2

(Solution):

$$\frac{\sqrt{0.64}-\sqrt{1.96}}{\sqrt{0.36}} + 1 = \frac{\sqrt{(0.8)^2}-\sqrt{(1.4)^2}}{\sqrt{(0.6)^2}} +1$$

$$=\frac{0.8-1.4}{0.6} +1$$

$$=\frac{-0.6}{0.6} +1=\text{-1+1=0}$$

-Answer C

23. $\sqrt{(0.6)^{-1}.6^{-1}} :(1.3)^{-1}=?$

A)3　　　B)2　　　C)$\frac{1}{2}$　　　D)$\frac{2}{3}$　　　E)$\frac{3}{8}$

(Solution):

$$\sqrt{(0.6)^{-1}.6^{-1}} :: (1.3)^{-1} = \sqrt{\left(\frac{6}{9}\right)^{-1}.\frac{1}{6}} : \left(\frac{13-1}{9}\right)^{-1}$$

$$= \sqrt{\frac{9}{6}.\frac{1}{6}} : \frac{9}{12}$$

$$\sqrt{\frac{9}{36}} : \frac{12}{9} = \frac{3}{6} : \frac{9}{12} = \frac{2}{3}$$

-Answer D

$$24. \sqrt{(-2)^{-4}} + \left(\frac{1}{3}\right)^{-1} - \sqrt[3]{-8} = ?$$

A)5.25   B)-1.25   C)5   D)1.25   E)9

(Solution):

$$\cdot\sqrt{(-2)^{-4}} + \left(\frac{1}{3}\right)^{-1} - \sqrt[3]{-8} = \sqrt{\frac{1}{(-2)^4}} + 3 - \sqrt[3]{-8}$$

$$\sqrt{\frac{1}{16}} + 3 - (-2)$$

$$= \frac{1}{4} + 3 + 2 = \frac{1}{4} + 5$$

$$=\frac{21}{4}=5.25$$

-Answer D

25. $\dfrac{\sqrt{7+\frac{1}{9}}}{0.13}$ =?

A)10   B)15   C)18   D)20   E)30

(Solution):

$$\frac{\sqrt{7+\frac{1}{9}}}{0.13} = \frac{\sqrt{\frac{64}{9}}}{\frac{13-1}{90}} = \frac{\sqrt{\frac{8}{3}}}{\frac{12}{90}}$$

$$=\frac{8}{3}\cdot\frac{90}{12} = 20$$

Yanit – Answer C

26. $\dfrac{2}{\sqrt[7]{8}}$.=?

A)$\sqrt[7]{2}$ B) $\sqrt[7]{4}$ C) $\sqrt[7]{8}$ D)$\sqrt[7]{16}$ E) $\sqrt[7]{32}$

(Solution):

$$\frac{2}{\sqrt[7]{8}} = \frac{2}{\sqrt[7]{2^3}} = \frac{2}{2^{\frac{3}{7}}} = 2^{1-\frac{3}{7}}$$

23

$$= 2^{\frac{4}{7}} = \sqrt[7]{2^4} = \sqrt[7]{16}$$

27. $\sqrt{2} \cdot \sqrt[3]{3} = ?$

A) $\sqrt[6]{6}$ B) $\sqrt[6]{36}$ C) $\sqrt[6]{48}$ D) $\sqrt[6]{54}$

E) $\sqrt[6]{54}$

(Solution):

$$\sqrt{2} \cdot \sqrt[3]{3} = \sqrt[6]{2^3} \cdot \sqrt[6]{3^2}$$

$$= \sqrt[6]{8} \cdot \sqrt[6]{9} = \sqrt[6]{72}$$

-Answer E

28. $\dfrac{\sqrt{2}}{\sqrt{2} + \dfrac{1}{\sqrt{2} + \dfrac{1}{\sqrt{2}}}} = ?$

A) $\dfrac{2}{3}$ B) $\dfrac{3}{4}$ C) $\dfrac{4}{5}$ D) $\dfrac{5}{4}$ E) $\dfrac{4}{3}$

(Solution):

24

$$\frac{\sqrt{2}}{\sqrt{2}+\frac{1}{\sqrt{2}+\frac{1}{\sqrt{2}}}}=\frac{\sqrt{2}}{\sqrt{2}+\frac{1}{\frac{3}{\sqrt{2}}}}=\frac{\sqrt{2}}{\sqrt{2}+\frac{\sqrt{2}}{3}}$$

$$=\frac{\sqrt{2}}{\frac{4\sqrt{3}}{3}}=\sqrt{2}.\frac{3}{4\sqrt{2}}=\frac{3}{4}$$

-Answer B

29. $\frac{\sqrt{(-4)^2}+3\sqrt{9}-\sqrt{(-3)^2}}{\sqrt{(-1)^2}+\sqrt{16}}=?$

A)1   B)2   C)3   D)4   E)5

(Solution):

$$\frac{\sqrt{(-4)^2}+3\sqrt{9}-\sqrt{(-3)^2}}{\sqrt{(-1)^2}+\sqrt{16}}=\frac{4+9-3}{1+4}$$

$$=\frac{10}{5}=2$$

-Answer B

$1\frac{\sqrt{20}+\sqrt{45}}{\sqrt{8}+\sqrt{18}}=?$

A) $\frac{3\sqrt{5}}{2}$ B) $2\sqrt{5}$ C) $\frac{\sqrt{10}}{2}$ D) $2\sqrt{10}$ E) $\frac{2\sqrt{10}}{3}$

(Solution):

$$\frac{\sqrt{20}+\sqrt{45}}{\sqrt{8}+\sqrt{18}} = \frac{2\sqrt{5}+3\sqrt{5}}{2\sqrt{2}+3\sqrt{2}} = \frac{5\sqrt{5}}{5\sqrt{2}} = \frac{\sqrt{5}}{\sqrt{2}} = \frac{\sqrt{10}}{2}$$

$-$ **Answer C**

2. $\frac{2\sqrt{3}}{\sqrt{2}} + \frac{3\sqrt{2}}{\sqrt{3}} = ?$

A) $2\sqrt{2}$ B) $2\sqrt{3}$ C) $3\sqrt{2}$ D) $2\sqrt{6}$ E) $2\sqrt{6}$

(Solution):

$$\frac{2\sqrt{3}}{\sqrt{2}} + \frac{3\sqrt{2}}{\sqrt{3}}$$

$$\frac{2\sqrt{3}}{\sqrt{2}} + \frac{3\sqrt{2}}{\sqrt{3}} = \frac{6}{\sqrt{6}} + \frac{6}{\sqrt{6}}$$

$$= \frac{12\sqrt{6}}{6}$$

$$= 2\sqrt{6}$$

3. $\sqrt{3^2} - \sqrt{(-3)^2} - (-2)(-3) = ?$

26

A)-6   B)0   C)3   D)6   E)12

(Solution):

$$\sqrt{3^2} = 3$$

$$\sqrt{(-3)^2} = |3| = 3$$

$$\sqrt{3^2} = \sqrt{(-3)^2}\text{-(-2).(-3)}$$

=3-3-(+6)=-6

-Answer A

4.$\dfrac{2^{1-n}.\sqrt{8^n}}{\sqrt{2^{-n}}}$=?

A)2^n B) 2^{n+1} C) 2^{-n}

D) $2^{\frac{1}{2}}$ E) 2^{-1}

(Solution):

$$\frac{2^{1-n}.\sqrt{8^n}}{\sqrt{2^{-n}}} = \frac{2^{1-n}.\sqrt{2^{3n}}}{2^{-\frac{n}{2}}}$$

$$\frac{2^{1-n}.2^{\frac{3n}{2}}}{2^{-\frac{n}{2}}}$$

$$2^{\left(1-n+\frac{3n}{2}+\frac{n}{2}\right)}=2^{n+1}$$

-Answer B

5. $\sqrt{2^2}.\sqrt{(-3)^2}-\sqrt{(-3)^2}-\sqrt{2^2}=?$

A)-5   **B)-3**   **C)-1**   **D)1**   **E)2**

(Solution):

$$\sqrt{2^2}.\sqrt{(-3)^2}-\sqrt{(-3)^2}-\sqrt{2^2}$$

$$\sqrt{2^2}=2$$

$$\sqrt{(-3)^2}=|-3|=3$$

$$=\sqrt{2^2}.\sqrt{(-3)^2}-\sqrt{(-3)^2}-\sqrt{2^2}$$

=2.3-3-2

=6-5 =1

-Answer D

6. $\sqrt{(-8)^2} - \sqrt[3]{(-8)^3} = ?$

A)-16   B)-8   C)0   D)8   E)18

(Solution):

$\sqrt{(-8)^2} - \sqrt[3]{(-8)^3}$

$\sqrt{(-8)^2} = |-8| = 8$

$\sqrt[3]{(-8)^3} = -8$

$\sqrt{(-8)^2} - \sqrt[3]{(-8)^3} = 8-(-8)=16$

        -Answer

7. $a=\sqrt{5}-1 \Rightarrow \left(\frac{1}{a} - \frac{1}{b}\right)^{\frac{1}{2}} = ?$

  $b=\sqrt{5}+1$

A)$\frac{\sqrt{2}}{2}$ B)$\frac{\sqrt{3}}{2}$ C)$2\sqrt{2}$ D)$2\sqrt{3}$ E)$4\sqrt{2}$

(Solution):

$$\left(\frac{1}{a} - \frac{1}{b}\right)^{\frac{1}{2}} = \left(\frac{1}{\sqrt{5}-1} - \frac{1}{\sqrt{5}+1}\right)^{\frac{1}{2}}$$

$$\frac{1}{\sqrt{5}-1} = \frac{1}{\sqrt{5}-1} \cdot \frac{\sqrt{5}+1}{\sqrt{5}+1} = \frac{\sqrt{5}+1}{5-1} = \frac{\sqrt{5}+1}{4}$$

$$= \frac{1}{\sqrt{5}+1} = \frac{1}{\sqrt{5}+1} \cdot \frac{\sqrt{5}-1}{\sqrt{5}-1} = \frac{\sqrt{5}-1}{5-1} = \frac{\sqrt{5}-1}{4}$$

$$= \left(\frac{1}{\sqrt{5}-1} - \frac{1}{\sqrt{5}+1}\right)^{\frac{1}{2}} = \left(\frac{\sqrt{5}+1}{4} - \frac{\sqrt{5}-1}{4}\right)^{\frac{1}{2}}$$

$$\left(\frac{\sqrt{5}+1-\sqrt{5}+1}{4}\right)^{\frac{1}{2}} = \left(\frac{2}{4}\right)^{\frac{1}{2}} = \sqrt{\frac{1}{2}} = \frac{1}{\sqrt{2}} = \frac{\sqrt{2}}{2}$$

-Answer B

8. $\dfrac{3}{\sqrt{7}-\sqrt{5}} \cdot \dfrac{3}{\sqrt{7}+\sqrt{5}} = 3p \Rightarrow p =?$

A)2   B)2   C)$\sqrt{2}$ D) $\sqrt{3}$ E) $\sqrt{5}$

(Solution):

$$\frac{3}{\sqrt{7}-\sqrt{5}} = \frac{3}{\sqrt{7}-\sqrt{5}} \cdot \frac{\sqrt{7}+\sqrt{5}}{\sqrt{7}+\sqrt{5}}$$

$$= \frac{3\sqrt{7}+3\sqrt{5}}{7-5} = \frac{3\sqrt{7}+3\sqrt{5}}{2}$$

$$\frac{3}{\sqrt{7}+\sqrt{5}} = \frac{3}{\sqrt{7}+\sqrt{5}} \cdot \frac{\sqrt{7}-\sqrt{5}}{\sqrt{7}-\sqrt{5}}$$

$$= \frac{3\sqrt{7}-3\sqrt{5}}{7-5} = \frac{3\sqrt{7}-3\sqrt{5}}{2}$$

$$\frac{3}{\sqrt{7}-\sqrt{5}}-\frac{3}{\sqrt{7}+\sqrt{5}}=\frac{3\sqrt{7}+3\sqrt{5}}{2}-\frac{3\sqrt{7}-3\sqrt{5}}{2}$$

$$\frac{3\sqrt{7}+3\sqrt{5}-3\sqrt{7}+3\sqrt{5}}{2}=$$

$$\frac{6\sqrt{5}}{2}=3\sqrt{5}=3p \Rightarrow p=\sqrt{5}$$

-Answer E

9. $\frac{\sqrt{0.81}+\sqrt{0.49}}{\sqrt{2.56}-\sqrt{1.44}}$ =?

A)0.4   B)0.2   C)1   D)2   E)4

(Solution):

$$\frac{\sqrt{0.81}+\sqrt{0.49}}{\sqrt{2.56}-\sqrt{1.44}}=\ =\frac{0.9+0.7}{1.6-1.2}$$

$$=\frac{1.6}{0.4}=4$$

-Answer E

10. $\sqrt{3+2\sqrt{2}}$ -$\sqrt{3-2\sqrt{2}}$ =?

A)$2\sqrt{3}$ B)$\sqrt{3}$ C)$\sqrt{2}$ D)3   E)2

(Solution):

$$\sqrt{3 + 2\sqrt{2}} - \sqrt{3 - 2\sqrt{2}} = \sqrt{(\sqrt{2} + 1)^2} - \sqrt{(\sqrt{2} - 1)^2}$$

$$= \sqrt{2} + 1 - (\sqrt{2} - 1)\sqrt{2} + 1 - \sqrt{2} + 1 = 2$$

-Answer E

11. $a, b \in Z$

$$\sqrt{72} - \sqrt{50} + \sqrt{27} = a\sqrt{2} + b\sqrt{3}$$

$$\Rightarrow 7a - b = ?$$

A)-3   B)-2   C)4   D)$2\sqrt{2}$ E)$3\sqrt{3}$

Çözü*m* (Solution):

$$\sqrt{72} - \sqrt{50} + \sqrt{27} = a\sqrt{2} + b\sqrt{3}$$

$$6\sqrt{2} - 5\sqrt{2} + 3\sqrt{3} = a\sqrt{2} + b\sqrt{3}$$

$$\sqrt{2} + 3\sqrt{3} = a\sqrt{2} + b\sqrt{3}$$

a=1$\Rightarrow$

b=3$\Rightarrow$ 7a-b=7-3=4

12. $\sqrt[3]{2} \cdot (3\sqrt[3]{32} - \sqrt[3]{108} + \frac{6}{\sqrt[3]{54}})$=?

A)$\sqrt[3]{4}$ B)$2\sqrt[3]{2}$ C)4   D)6   E)8

(Solution):

$\sqrt[3]{2} \cdot (3\sqrt[3]{32} - \sqrt[3]{108} + \frac{6}{\sqrt[3]{54}})$

$=\sqrt[3]{2} \cdot (3 \cdot 2\sqrt[3]{4} - 3\sqrt[3]{4} + \frac{6}{3\sqrt[3]{2}})$

$=\sqrt[3]{2} \frac{\cdot(3 \cdot 2\sqrt[3]{4} - 3\sqrt[3]{4}+6)}{3\sqrt[3]{2}}$

$=\frac{18 \cdot 2 - 9 \cdot 2 + 6}{3}$

$=\frac{18+6}{3}$

$=\frac{24}{3}=8$ -Answer E

WORKBOOK TESTS

1.$5^{X+1}=\sqrt{25^{3X}} \Rightarrow X =?$

A)0   B)1   C)$\frac{1}{2}$ D)$\frac{2}{3}$ E)4

2.$d^2 =\sqrt{2^{n+2}} \Rightarrow \sqrt[3]{d^6}=?$

A)2^n B)$2^{\frac{1}{2}+n}$ C)$2^{\frac{n+2}{2}}$ D)2^{n-2} E)2^{3n+6}

3.$\sqrt{2010.1998 + 36} =?$

A)1997   B)1999   C)2000

D)2002   E)2004

4.$\sqrt{x + \sqrt{x}}+\sqrt{x - \sqrt{x}} = 4 \Rightarrow x =?$

A.$\frac{9}{4}$ B)$\frac{21}{8}$ C)$\frac{36}{11}$ D)$\frac{64}{15}$ E)$\frac{72}{18}$

5.x.$\sqrt{\frac{4}{3}} = \sqrt{\frac{3}{4}}+\sqrt{\frac{4}{3}} \Rightarrow x =?$

A.$\frac{1}{2}$ B)$\frac{3}{5}$ C)$\frac{6}{7}$ D)$\frac{7}{4}$ E)$\frac{8}{5}$

6.$\sqrt{3.\sqrt[3]{3^x}}=\frac{1}{243}$ $\Rightarrow x =?$

A)-3   B)-14   C)-16   D)-18

E)-33

7.$\sqrt{x+\sqrt{x^2}}.\sqrt[3]{x+\sqrt{x^2}} = 16^{\frac{5}{12}} \Rightarrow x =?$

A)0   B)1   C)2   D)3   E)4

8.$\frac{6-\sqrt{6}}{\sqrt{3}-\sqrt{2}}=?$

A)$2\sqrt{3}-2$ B) $4\sqrt{3}+3\sqrt{2}$ C)$2\sqrt{2}-\sqrt{3}$ D)) $\sqrt{3}+\sqrt{2}$

E)) $\sqrt{2}-3\sqrt{3}$

9.$\sqrt[x]{3^x\sqrt{729}} = 3 \Rightarrow X =?$

A)0   B)1   C)2   D)3   E)4

10. $X=\sqrt[3]{\dfrac{2}{\sqrt[3]{2}}} \Rightarrow x^{18} = ?$

A)8   B)16   C)21   D)32   E)64

11. $\sqrt{\dfrac{3^{X+2}}{9^{X-1}}}=27 \Rightarrow X = ?$

A)-5   B)-2   C)2   D)3   E)8

12. $\dfrac{(\sqrt{5}-2).\left(\sqrt{9+2\sqrt{20}}\right)}{\sqrt{2}}=?$

A) $\dfrac{-\sqrt{2}}{3}$ B)1   C) $\dfrac{\sqrt{2}}{2}$ D) $\dfrac{-3}{2}$ E) $\dfrac{\sqrt{3}}{5}$

13. $\sqrt[4]{27\sqrt[4]{27\sqrt[4]{27}}} = X,$

$$\sqrt{5\sqrt{5\sqrt{5}}} = Y \Rightarrow Y^2 - x^2 = ?$$

A)8   B)12   C)16   D)21   E)27

14. $\frac{1}{2-3\sqrt{3}} + \frac{1}{2+3\sqrt{3}} = ?$

A)$-\frac{2}{3}$ B)$\frac{-14}{5}$ C)$\frac{-4}{23}$

D)$\frac{-8}{17}$ E)$\frac{\sqrt{3}}{2}$

15. $\frac{5}{5-\sqrt{5}} \cdot (5+\sqrt{5})^{-1} = ?$

A)$\frac{1}{2}$ B)$\frac{5}{2}$ C)$\frac{1}{4}$ D)$\frac{4}{7}$ $\frac{7}{2}$

16. $\sqrt{9} + \sqrt{4} - \sqrt{(-4)^2} - \sqrt{(-2)^2} = ?$

A)1   B)11   C)-10   D)-11

E)-5

17. $\dfrac{1}{\sqrt{7-4\sqrt{3}}} + \dfrac{1}{2+\sqrt{3}} = ?$

A)2   B)$2\sqrt{3}$ C)1   D)4   E)$\sqrt{3}$

18. $\sqrt{0.16} + \sqrt{0.64} = ?$

A)$\dfrac{3}{2}$ B)$\dfrac{6}{5}$ C)$\dfrac{4}{5}$

D)$\dfrac{9}{10}$ E)$\dfrac{12}{7}$

19. $\sqrt{\dfrac{3^{-1}}{0.3} : \dfrac{0.09}{10}} = ?$

A)$\dfrac{10}{3}$ B)$\dfrac{10}{9}$ C)$(\dfrac{9}{10})^{-2}$

D)$\dfrac{3}{10})^{-2}$ E)$\dfrac{3}{5}$

20. $\sqrt{6 + \sqrt{6 + \sqrt{6 + \sqrt{6 + \ldots \ldots \ldots}}}} \cdot \sqrt{X + \sqrt{X + \sqrt{X + \ldots \ldots \ldots}}} \Rightarrow$

$$\Rightarrow n = ?$$

A)8　　　　B)12　　　　C)16　　　D)27　　　E)64

21. $\frac{1}{\sqrt{2}-1} - \frac{1}{1+\sqrt{2}} = ?$

A)0　　　B)$\sqrt{2}$　　　C)1　　　　D)-1　　　E)2

22. $3^X=8 \Rightarrow (9^X)^{-1} \cdot (81^X)^2 = ?$

A)a^{-2}　　　　　B)a^2　　　　　　C) a^4　　　　　D) a^{-5}
E) a^5

23. $\sqrt{a} + \frac{1}{\sqrt{a}} = \sqrt{4} \Rightarrow a^2 + \frac{1}{a^2} = ?$

A)9　　　　　B)8　　　　C)4　　　　D)2　　　　E)1

24. $\sqrt{3^4\sqrt{3^{2X}}} = (\frac{1}{81})^2 \Rightarrow X = ?$

A)-27　　　　　B)81　　　　C)30　　　　D)36　　　E)-21

(Answers)

1.C   2.C   3.E   4.D   5.D   6.E

7.C   8.B   9.D   10.B   11.B   12.C

13.C   14.C   15.C   16.A   17.D   18.B

19.D   20.D   21.E   22.E   23.D   24.C

$1.\ 4\sqrt{8} + 5\sqrt{18} - 3\sqrt{72} + \sqrt{50} = ?$

A)$6\sqrt{2}$ B) $7\sqrt{2}$ C) $8\sqrt{2}$ D) $9\sqrt{2}$

E) $10\sqrt{2}$

$2.\ \sqrt{108} - \sqrt{48} - \sqrt{75} = ?$

A)$\sqrt{3}$ B)$-3\sqrt{3}$ **C)** $2\sqrt{3}$ **D)**$-\sqrt{3}$
E)0

3. $3\sqrt[3]{2} + 4\sqrt[3]{16} - 4\sqrt[3]{54} = ?$

A)-2$\sqrt[3]{2}$ **B)-$\sqrt[3]{3}$**   **C)-$\sqrt[3]{2}$**   **D)$\sqrt[3]{3}$**

E)2$\sqrt[3]{3}$

4.$\sqrt[3]{0.006} + \sqrt[3]{0.002} = ?$

A)$\dfrac{\sqrt[3]{2}(\sqrt[3]{2}+1)}{10}$ B)$\dfrac{\sqrt[3]{2}(\sqrt[3]{3}+1)}{10}$ C)$\dfrac{\sqrt[3]{3}(\sqrt[3]{2}-1)}{10}$

D) $\frac{\sqrt[3]{3}(\sqrt[3]{2}+1)}{10}$ E)$\sqrt[3]{5} + \sqrt{2}$

5. $\sqrt{1 - \frac{9}{25}} + \sqrt{1 - \frac{11}{36}}$ =?

A)$\frac{37}{25}$ B)$\frac{27}{25}$ C)$\frac{49}{30}$

D) $\frac{51}{25}$

E) $\frac{49}{16}$

6.$\sqrt{4^2 - 3^2}$-$\sqrt[4]{7^2}$ =?

A)0   B)1   C)2   D)3   E)40

7.$\frac{\sqrt[3]{16}+\sqrt[3]{54}-\sqrt[3]{250}+\sqrt[3]{128}}{\sqrt[3]{16}-\sqrt[3]{250}}$=?

A)$-\frac{1}{2}$ B)$-\frac{2}{3}$ C)$-\frac{3}{4}$

D)$-\frac{4}{3}$ E)$-\frac{5}{4}$

8. $3\sqrt{2} + 4\sqrt{8} - 5\sqrt{50} + 8\sqrt{32}$ =?

A)$10\sqrt{2}$ B) $12\sqrt{2}$ C) $4\sqrt{3}$

D) $16\sqrt{2}$ E) $18\sqrt{2}$

9. $2\sqrt[3]{3} - 3\sqrt[3]{24} + 4\sqrt[3]{81}$ =?

A)$\sqrt[3]{3}$ B)$2\sqrt[3]{3}$ C)$4\sqrt[3]{3}$

D)$6\sqrt[3]{3}$ E) $8\sqrt[3]{3}$

10. $X > 0, Y > 0, Z > 0 \Rightarrow$

$6\sqrt{XY^2Z^2} + \frac{8}{2}\sqrt{XY^2Z^4} - \frac{6}{Y}\sqrt{XY^4Z^2}$ =?

A)$8YZ\sqrt{X}$ B) $6YZ\sqrt{X}$ C) $\sqrt{X^2-1}$

D) $\sqrt{X-1}$ E)$6Y\sqrt{X-1}$

11. $X > 1 \Rightarrow \sqrt{(X+1)^3} - X\sqrt{X+1} + \sqrt{(X-1)(X^2-1)}$ =?

43

A) $X\sqrt{X+1}$ B) $\sqrt{X+1}$ C)$\sqrt{X^2-1}$

D) $\sqrt{X-1}$ E) $X\sqrt{X-1}$

12. $\sqrt{\dfrac{3}{2}} + \sqrt{\dfrac{2}{3}} = ?$

A)$\dfrac{\sqrt{3}}{6}$ B)$6\sqrt{\dfrac{6}{5}}$ C)$3\dfrac{\sqrt{6}}{2}$

D)$\dfrac{5}{\sqrt{6}}$ E)$2\sqrt{6}$

13. $\sqrt{0.04} - 2\sqrt[3]{0.008} - \sqrt[4]{0.0016} + \sqrt{1.69} = ?$

A)0.5   B)0.6   C)0.7   D)0.8

E)0.9

14. $\dfrac{1}{\sqrt[3]{a^2}} = ?$

A)$\sqrt[3]{a^2}$ B)$\sqrt[3]{a}$ C)$\dfrac{\sqrt[3]{a}}{a}$

D)$\dfrac{\sqrt[3]{a^2}}{a}$ E)$\sqrt{a}$

15. $\sqrt{a}.\sqrt[3]{b}.\dfrac{1}{\sqrt[3]{ab}}=?$

A)1   B) $\sqrt{a}$ C) $\sqrt[3]{ab}$

D) $\sqrt[6]{a}$ E) $\sqrt[6]{ab}$

16. $\sqrt{2\sqrt[3]{2\sqrt{2}}}=?$

A) $\sqrt[12]{2^8}$ B) $\sqrt[6]{2^5}$ C) $\sqrt[12]{2^5}$

D) $\sqrt[4]{2^3}$ E) $\sqrt[16]{2^8}$

17. $\sqrt[4]{\dfrac{x^3}{\sqrt[3]{x^2}}}:x^{\frac{7}{12}}=?$

A)0   B)1   C)2   D)3   E)x^4

18. $\dfrac{\sqrt[3]{5}.\sqrt{2}}{\sqrt[6]{10}}=?$

A) $\sqrt[6]{10}$ B) $\sqrt[5]{15}$ C) $\sqrt[5]{20}$

D) $\sqrt[6]{25}$ E) $\sqrt[5]{200}$

19. $\sqrt[2]{2} \cdot \sqrt[3]{2} \cdot \sqrt{2} = ?$

A) $2\sqrt[3]{5}$ B) $2\sqrt[30]{2}$ C) $2\sqrt[15]{3}$

D) $2\sqrt[30]{7}$ E) $2\sqrt[30]{3}$

20. $\sqrt[4]{2\sqrt[3]{2\sqrt{2}}} = ?$

A) $\sqrt[16]{8}$ B) $\sqrt[4]{4}$ C) $\sqrt[8]{8}$

D) $\sqrt[24]{8}$ E) $\sqrt[12]{6}$

21. $\dfrac{\sqrt{300}-2\sqrt{27}}{\sqrt{75}+\sqrt{3}} = ?$

A)1 B)$\frac{1}{2}$ C)$\frac{2}{3}$

D)$2\sqrt{3}$ E)$3\sqrt{3}$

(Answers)

1.E 2.B 3.C 4.B 5.C 6.A

7.D 8.E 9.E 10.A 11.A 12.D

13.E 14.C 15.D 16.D 17.B 18.C

19.B 20.C 21.B

1. $\sqrt{\sqrt{0.0036} + \sqrt{0.09}} \cdot \frac{10}{\sqrt{2}} = ?$

A)3   B)$5\sqrt{2}$ $\sqrt{3}$

D) $3\sqrt{2}$ E)5

2. $\frac{\sqrt[6]{(64)^{-1}}}{\sqrt[3]{4} \cdot \sqrt[4]{4}} \cdot \sqrt[6]{2} = ?$

A)$\frac{1}{2}$ B)2   C)8   D)$\frac{1}{4}$

E) $\sqrt[6]{2}$

3. $\frac{\sqrt{32} - \sqrt{45} + \sqrt{2} - \sqrt{20}}{\sqrt{5} - \sqrt{2}} = ?$

A)-5   B)$\sqrt{5} + \sqrt{2}$ C)4

D)$\sqrt{2}$ E)$3\sqrt{5}$

4. $\frac{a}{\sqrt[n]{a^{n-1}}} = ?$

A)$\sqrt[n]{a}$ B)$\sqrt[n]{a^1}$ C) $\sqrt[n]{a^{n+1}}$

D) $\sqrt[1]{a}$ E) $\sqrt[n-1]{a}$

5.$\sqrt{48} - \sqrt{12} - \sqrt{\frac{4}{3}}\cdot\sqrt{\frac{1}{3}} = ?$

A)0   B)1   C)$\sqrt{3}$

D)$\frac{1}{3}\sqrt{3}$ E)$\frac{2}{3}\sqrt{3}$

6.$\sqrt{1.21} - 35.\sqrt[3]{0.008} - 10.\sqrt[4]{\frac{0.0016}{(0.25)^2}} = ?$

A)0   B)$\frac{1}{2}$ C)$\frac{2}{3}$ D)5   E)11

7.$2-\left[3:\left(\sqrt{3:2}\right)^{-2} - 2:\left(\sqrt{2:3}\right)^{-2}\right]-3=?$

A)-3   B)-2   C)0

D)$\frac{2}{3}$ E) $\frac{3}{2}$

8. $(\sqrt[9]{\sqrt[3]{27x^6}})^2-.(\sqrt[6]{\sqrt[4]{x^8}})^2=?$

A) X   B)$2\sqrt{X}$ C) $3\sqrt{X}$

D) X$\sqrt{X}$ E)$2x^2$

9. $\sqrt{18} - \sqrt[3]{16} - 3\sqrt{2} + \sqrt[3]{54} = a^3\sqrt{2} =?$

A)-1 B) 0   C) 1

D) $\sqrt[3]{2}$ E) $\sqrt{2}$

10. $\dfrac{5\sqrt{\frac{1}{2}} - \sqrt{0.5} + \sqrt{200}}{\sqrt{8}} =?$

A)12   B) $6\sqrt{2}$ C)6

D)4   E) $2\sqrt{2}$

11 . $4\sqrt{45} - 2\sqrt{80} - \sqrt[4]{25} =?$

$A)3\sqrt{5}$ B) $2\sqrt{5}$ C) $\sqrt{5}$
$D)0$ E) $9\sqrt{5}$

12. $\sqrt{45} - 10\sqrt{\frac{1}{5}} + \sqrt{80} - \sqrt[4]{25}=?$

A) $3\sqrt{5}$ B) $7\sqrt{5}$ C) $4\sqrt{5}$
D) $5\sqrt{5}$ E) $2\sqrt{5}$

13. $3\sqrt{\frac{4}{3}} - 2\sqrt{\frac{25}{3}} + \sqrt{\frac{49}{3}} =?$

A) $\sqrt{3}$ B) 2 $\sqrt{3}$ C) 3 $\sqrt{3}$

D) $\frac{\sqrt{3}}{3}$ E) 5 $\sqrt{3}$

14. $\frac{\sqrt{5.76}+\sqrt{2.89}+\sqrt{1.96}}{\sqrt{0.49}+\sqrt{0.15}} = ?$

A) $\frac{1}{2}$ B) 2   C) 3

D) 4   E) 5

15. $\sqrt{8.1} + \sqrt{4.9} - \sqrt{12.1} = ?$

A) $\sqrt{10}$ B) 2 $\sqrt{10}$ C) $\frac{\sqrt{10}}{2}$

D) $\frac{\sqrt{10}}{5}$ E) 5

16. $\sqrt{\frac{4}{25} - \frac{3}{5} + \frac{9}{16}} = ?$

A) -1   B) 2   C) $\frac{1}{3}$

51

D) $\frac{7}{20}$ E) $\frac{1}{4}$

17. $\dfrac{7}{\sqrt{7}-\dfrac{3}{\sqrt{7}-\dfrac{3}{\sqrt{7}}}}=?$

A)1   B) $\frac{1}{4}$ C)$4\sqrt{7}$

D)4   E) $2\sqrt{7}$

18. $3\sqrt{0.64}+8\sqrt{0.49}=?$

A)4   B)5   C)6   D)7   E)8

19. $\sqrt[3]{\dfrac{6}{7^{1-3X}}+\dfrac{7^{3X}}{7}}=?$

A)7^{2X} B) 7^{3X} C) 7^{X}

D)7   E)49

20. $\dfrac{\sqrt{252}}{\sqrt{7}}+\dfrac{\sqrt{27}}{\sqrt{\frac{1}{3}}}=?$

A)1　　　　B)3　　　　C)6

D)9　　　　E)15

$21\sqrt{0.21 + \sqrt{0.0016}} + \sqrt{0.53 - \sqrt{0.000064}} = ?$

A)1　　　B)0.3　　　C)1.1

D)1.2　　E)1.5

$22.4\sqrt{2.52} - 2\sqrt{3.43} = ?$

A)$\sqrt{7}$　　　　B)$2\sqrt{7}$　　　C) $3\sqrt{7}$

D)0　　　　E)1

$23.\sqrt[5]{0.008} : \sqrt[5]{25} = ?$

A)1　　　B)0.1　　C)0.2　　D.0.6　　E)$\frac{1}{50}$

$24.\sqrt[3]{0.5} . \sqrt[6]{0.25} . \sqrt[12]{0.0625} = ?$

A)1   B)$\frac{1}{2}$ C)$\frac{1}{4}$ D)5   E)$\frac{1}{5}$

(Answers)

1.D   2.D   3.A   4.B   5.C   6.A

7.C   8.E   9.C   10.C   11.A   12.C

13.A   14.E   15.C   16.D   17.C   18.E

19.C   20.E   21.D   22.A   23.C   24.B

1. $(4\sqrt{5} - 2\sqrt{3})^2 - (4\sqrt{5} + 2\sqrt{3})^2 = ?$

A)$32\sqrt{15}$ B) $16\sqrt{15}$ C)$-20\sqrt{15}$

D)$-24\sqrt{15}$ E)$-32\sqrt{15}$

2. $\sqrt{6 - \sqrt{6 - \sqrt{6 - \sqrt{6 \ldots \ldots}}}} = ?$

A)1   B)2   C)3   D)6   E)36

3. $\sqrt{3\sqrt{3\sqrt{3} \ldots \ldots}} = a \Rightarrow a^2 = ?$

A)3   B)6   C)9   D)12   E)36

4. $\sqrt[3]{49 : \sqrt[3]{49 : \sqrt[3]{49 \ldots \ldots}}} X$

$\sqrt{4\sqrt{4\sqrt{4} \ldots \ldots}} = Y \Rightarrow x^2 - Y = ?$

A)3   B)4   C)7   D)45   E)53

$$5. \sqrt{6 + \sqrt{6 + \sqrt{6 \ldots \ldots \ldots}}} = ?$$

A-2   B)-3   C)0

D)2   E)3

$$6. \sqrt{7 - 2\sqrt{12}} + \sqrt{8 + 2\sqrt{15}} = ?$$

A)$2+\sqrt{5}$ B) $2-\sqrt{5}$ C) $\sqrt{5}+1$ D) $2\sqrt{3}$

E) $\sqrt{3}$

$$7. \sqrt{5 + 2\sqrt{6}} + \sqrt{8 - 2\sqrt{15}} - \sqrt{9 - 4\sqrt{5}} = ?$$

A) $3+\sqrt{3}$ B) $5\sqrt{2}$ C) $2+\sqrt{2}$

D) $4+\sqrt{5}$ E) $2\sqrt{3}$

$$8. \frac{4}{\sqrt{10+2\sqrt{21}}} + \sqrt{3} = ?$$

A) $4\sqrt{3}$ B) $5\sqrt{2}$ C) $6\sqrt{7}$

D) $\sqrt{7}$ E) $2\sqrt{3}$

9. $\sqrt{5 + 2\sqrt{4}} \cdot \sqrt{5 - 2\sqrt{4}} = ?$

A)2   B)3   C)4

D)9   E)16

10. $\dfrac{1}{\sqrt{6-2\sqrt{8}}} - \dfrac{1}{\sqrt{6+2\sqrt{8}}} = ?$

A)$\sqrt{2}$ B) $\sqrt{3}$ C)$2\sqrt{3}$

D) $2\sqrt{2}$ E) $3\sqrt{2}$

11. $\sqrt{\dfrac{\sqrt{3}+1}{\sqrt{3}-1}} - \cdot\sqrt{\dfrac{\sqrt{3}-1}{\sqrt{3}+1}} = ?$

A. $2\sqrt{3}$ B)$-\sqrt{3}$ C)$-2\sqrt{3}$

D) $\sqrt{2}$ E)3

12. $\dfrac{\sqrt{3}+\sqrt{5}}{\sqrt{3}-\sqrt{5}} - \dfrac{\sqrt{3}-\sqrt{5}}{\sqrt{3}+\sqrt{5}} = ?$

A) $-\sqrt{15}$ **B)** $-2\sqrt{15}$ **C)** $-3\sqrt{15}$

D) $-4\sqrt{15}$ **E)** $\sqrt{15}$

13. $\sqrt{2\sqrt{\dfrac{1}{X}}} = 2 \Rightarrow X = ?$

A) 2^{-6} **B)** 2^{-5} **C)** 2^{-4}

D) 2^{4} **E)** 2^{5}

14. $\dfrac{\sqrt[8]{16}.\sqrt[4]{0.25}}{\sqrt{0.1}} = ?$

A) 10   **B)** $2\sqrt{5}$ **C)** $\sqrt{6}$

D) $\sqrt{10}$ **E)** $4\sqrt{5}$

15. $\sqrt[a]{\dfrac{9^{a+1}-3^{2a}}{8,3^{a}}} = ?$

A) 1   **B)** 2   **C)** 3

D)3^a E) 3^{-a}

16. $\sqrt[8]{2\sqrt{2}} = a \Rightarrow \sqrt[3]{a^{16}} = ?$

A)1   B)2   C)4

D)8   E)16

17. $\dfrac{1}{\sqrt{0.08}} + \dfrac{2}{\sqrt{0.32}} = ?$

A)1   B)$\sqrt{2}$ C)10

D)$5\sqrt{2}$ E)$2\sqrt{3}$

18. $\sqrt{3\sqrt[3]{X}} = \sqrt[3]{2\sqrt{3}} \Rightarrow X = ?$

A)3   B)6   C)27

D)$\dfrac{4}{9}$ E)$\dfrac{3}{2}$

19. $\dfrac{\sqrt{6X} - \sqrt{3X}}{\sqrt{2} - 2} = -\sqrt{6} \Rightarrow X?$

A)$\sqrt{2}$ B)$\sqrt{5}$ C)4

D)2   E)$\sqrt{22}$

20.$(\sqrt{3}+\sqrt{2})^{-100} \cdot (5-\sqrt{24})^{-50} =?$

A)-1   B)1   C)3^{40}

D) 2^{100} E) $(\sqrt{3}+\sqrt{2})^{-50}$

21.$\sqrt[3]{5\sqrt[4]{5}\sqrt[3]{5}}=5^X \Rightarrow X =?$

A)$\frac{4}{9}$ B) $\frac{4}{5}$ C) $\frac{8}{9}$

D) $\frac{6}{5}$ E)2

22.$\sqrt{8-27-\sqrt{48}} =?$

A)$\sqrt{3}-1$ B) $\sqrt{3}+1$ C)$\sqrt{3}+\sqrt{2}$

D) $\sqrt{3}+2$ E)2 $\sqrt{3}+2$

(Answers)

1.A	2.D	3.A	4.B	5.B	6.E
7.E	8.C	9.C	10.D	11.E	12.A
13.C	14.D	15.C	16.B	17.D	18.D
19.C	20.E	21.A	22.B		

13. $\dfrac{\sqrt{36}}{\sqrt[4]{81}}=?$

A)5   B)4   C)3   D)2   E)1

(Solution)

$$\frac{\sqrt{36}}{\sqrt[4]{81}} = \frac{\sqrt{6^2}}{\sqrt{3^4}} = \frac{6}{3} = 2$$

-Answer D

14. $\dfrac{\sqrt[3]{128}+\sqrt[3]{16}}{\sqrt[3]{2}} =?$

A)$\sqrt[3]{2}$ B)$2\sqrt[3]{2}$ C)$3\sqrt[3]{2}$ D)3   E)6

(Solution)

$$\frac{\sqrt[3]{128}+\sqrt[3]{16}}{\sqrt[3]{2}} = \frac{\sqrt[3]{2^6}.\sqrt[3]{2}+\sqrt[3]{2^3}.\sqrt[3]{2}}{\sqrt[3]{2}}$$

$$\frac{4\sqrt[3]{2}+2\sqrt[3]{2}}{\sqrt[3]{2}} = 6\frac{\sqrt[3]{2}}{\sqrt[3]{2}} = 6$$

-Answer E

15. $x^a=\sqrt{5} \Rightarrow x^{-4a} =?$

A) $\frac{1}{125}$ B) $\frac{1}{25}$ C) $\frac{1}{5}$ D)5   E)25

Cozum(Solution)

$x^a=\sqrt{5} \Rightarrow x^{-4a}=(x^a)^{-4}$

$(\sqrt{5})^{-4}$

$\frac{1}{(\sqrt{5})^{-4}} = \frac{1}{25}$

-Answer B

16. $\frac{\sqrt{10}}{\sqrt{2}+\sqrt{6}} + \frac{\sqrt{10}}{\sqrt{6}-\sqrt{2}} =?$

A) $\sqrt{2}$ B) $\sqrt{3}$ C) $\sqrt{5}$ D) $\sqrt{15}$ E)2 $\sqrt{15}$

(Solution)

$\frac{\sqrt{10}}{\sqrt{2}+\sqrt{6}} + \frac{\sqrt{10}}{\sqrt{6}-\sqrt{2}}$

$(\sqrt{5} - \sqrt{2}) \quad (\sqrt{5} + \sqrt{2})$

$=\frac{\sqrt{60}+2\sqrt{10}+\sqrt{60}-2\sqrt{10}}{6-2}$

$=\frac{2\sqrt{60}}{4}$

$$=\frac{8\sqrt{15}}{4} = 2\sqrt{15}$$

-Answer E

17. $\sqrt{-X + 2\sqrt{X-1}} + \sqrt{Y - \sqrt{2Y-1}} = 0 \Rightarrow X + Y = ?$

A)2   B)3   C)4   D)5   E)6

(Solution)

$$\sqrt{-X + 2\sqrt{X-1}} + \sqrt{Y - \sqrt{2Y-1}} = 0$$

$\Rightarrow -X + 2\sqrt{X-1} = 0$ $Y - \sqrt{2Y-1} = 0$

$2\sqrt{X-1} = 0$ $Y = \sqrt{2Y-1}$

$X = 2\sqrt{X-1}$ $(Y)^2 = (\sqrt{2Y-1})^2$

$(X)^2 = \left(2\sqrt{X-1}\right)^2$ $Y^2 = 2Y - 1$

$X^2 = 4X - 4$ $Y^2 - 2Y + 1 = 0$

$X^2 - 4X + 4 = 0$ $(Y-1)^2 = 0$

$\Rightarrow X = 2 \Rightarrow X + Y = 2 + 1 = 3$ $\Rightarrow Y = 1$

1. $\dfrac{\sqrt{aa}.\sqrt{bb}}{\sqrt{a}\sqrt{b}} =?$

A)$11\sqrt{a}$ B)11b C)10

D)$\sqrt{a}$ E)11

2. $a,b \in R$

$a.b=16 \Rightarrow \sqrt[4]{a\sqrt{b}}.\sqrt[4]{b\sqrt{a}} =?$

A)$\sqrt{2}$ B)$2\sqrt{2}$ C)$2+\sqrt{3}$

D) $4+\sqrt{3}$ E)8

3. $\dfrac{\sqrt{3}-2}{1+\sqrt{2}} = P \Rightarrow \dfrac{1-\sqrt{2}}{2+\sqrt{3}} =?$

A)p B)$\sqrt{3}P$ C)2P

D)3P E)4P

4. $\sqrt{2\sqrt[3]{X}} = 2\sqrt{2} \Rightarrow X =?$

A)4   B)8   C)16

D)32   E)64

5)$\frac{X-\sqrt{45}+20}{\sqrt{180}-X} = 4 \Rightarrow X =?$

A)$8\sqrt{5}$ B) $5\sqrt{5}$ C) $2\sqrt{8}$

D) $\sqrt{5}$ E)2

6.$\frac{\sqrt[3]{(-4)^3}}{\sqrt{(-4)^2}}+\frac{\sqrt{(-7)^2}}{\sqrt{49}} =?$

A)0   B)1   C)2   D)3   E)6

7.$3^X + 3^{X-2} = 30 \Rightarrow \sqrt[X]{0.125} =?$

A)$\frac{1}{5}$ B)$\frac{\sqrt[3]{3}}{5}$ C)$\frac{1}{2}$

D))$\frac{1}{8}$ E)1

8.$\sqrt[4]{3\sqrt[3]{X}} = \sqrt[4]{27\sqrt[3]{3}} \Rightarrow X =?$

A)$3^{\frac{1}{12}}$ B)$3^{\frac{1}{3}}$ C)3^3

D) 3^8 E)3

9.$5^a = x$

$\sqrt{x^2\sqrt{x}} = 25 \Rightarrow a =?$

A)$\frac{1}{5}$ B)1   C)$\frac{8}{5}$

D)2   E)8

10.$(5\sqrt{2} + 2\sqrt{3})^2 = X + Y\sqrt{150} \Rightarrow X + Y =?$

A)16   B)32   C)48   D)54

E)66

11.$\sqrt{36X + 36} + \sqrt{9X + 9} = 18 \Rightarrow X =?$

A)2   B)3   C)4

D)5   E)6

12. $\sqrt[n]{16^6 + 8^8} \in z \Rightarrow \min(n) =?$

A)16   B)20   C)24

D)25   E)10

13. $\sqrt[4]{\dfrac{1}{81} + \dfrac{1}{144} - \dfrac{1}{54}}?$

A)$\dfrac{2}{3}$ B)$\dfrac{1}{3}$ C)$\dfrac{1}{6}$

D)$\dfrac{1}{9}$ E)$\dfrac{1}{2}$

14. $\sqrt{2^{X+3} + 2^X} = 48 \Rightarrow X =?$

A)4   B)5   C)6

D)7   E)8

15. $X + a = \sqrt{a^2 + 6} \Rightarrow$ x.y=?

$\quad$ y-a= $\sqrt{a^2 + 6}$

A)6   B)9   C)12

D)15   E)24

16. $\sqrt{4^{X-1}} \cdot \sqrt{2^X} = 16 \Rightarrow X = ?$

A)$\sqrt[3]{2}$ B)2   C)4

D)$\sqrt{2}$ E)$2\sqrt{2}$

17. $\sqrt[3]{a\sqrt{a}} = 2 \Rightarrow \sqrt{a\sqrt{a}} = ?$

A)$\sqrt{2}$ B)3   C) $2\sqrt{2}$

D)4   E) $3\sqrt{2}$

18. $\sqrt{22.5} + \sqrt{8.1} = a\sqrt{10} \Rightarrow a = ?$

A)$\frac{12}{5}$ B)$\frac{11}{5}$ C)$\frac{3}{5}$

D)2   E)3

19. $\left(\sqrt{2} - 1\right)^2 \cdot \left(\sqrt{6} + \sqrt{3}\right)^2 = ?$

A)3   B)2   C)1

D)$\sqrt{2}$ E)$\sqrt{3}$

20. $\dfrac{a}{b} - \dfrac{b}{a} = 4 \Rightarrow \sqrt{\dfrac{a^4+b^4}{a^2b^2} + 46} = ?$

A)6   B)7   C)8

D)9   E)10

21. $\sqrt{\dfrac{15}{4^{1-a}} + 4^{a-1}} = 64 \Rightarrow a = ?$

A)2   B)3   C)4   D)5   E)6

22. X+Y=0 $\Rightarrow \left(X + \sqrt{X^2 + 1}\right).\left(Y + \sqrt{Y^2 + 1}\right) = ?$

A)1   B)$x^2 - 1$ C)$2\,x^2 - 1$

D)-1   E) $2x^2$

23, X,Y$\in R$

$\sqrt{X - 2Y} + \sqrt{Y + 2} = 0 \Rightarrow Z = ?$

X+Y-Z=8

A)-16   B)-14   C)-12

D)-6   E)-3

1. $\dfrac{\sqrt{6}+\sqrt{2}}{\sqrt{6}-\sqrt{3}+\sqrt{2}-1}\cdot\dfrac{2}{\sqrt{2}}=?$

A)2   B)$\sqrt{2}$ C) $1+\sqrt{2}$

D)$2+\sqrt{2}$ E)4

2. $\sqrt{(1+\sqrt{5})^{2}}\cdot\sqrt{6-2\sqrt{5}}=?$

A)$-\sqrt{5}$ B)-1   C)1

D)14   E)9

3.X,Y$\in R$

$\sqrt{\dfrac{X}{Y}}+\sqrt{\dfrac{Y}{X}}=3X+3Y \Rightarrow X.Y=?$

A)$\dfrac{1}{9}$ B)$\dfrac{1}{3}$ C)1

D)3   E)9

4. $\sqrt[3]{24 + \sqrt[3]{X\sqrt[3]{8}}} = 3 \Rightarrow X =?$

A)21   B)25   C)27

    D)34   E)40

5. $\dfrac{\sqrt[3]{4^{X+1}}}{\sqrt[3]{8^{X-1}}} = 16 \Rightarrow X =?$

A)$-\dfrac{13}{5}$ B)$-\dfrac{11}{5}$ C)$-\dfrac{9}{5}$

   D)$\dfrac{11}{5}$ E)$\dfrac{13}{5}$

6. $\sqrt[X]{0.16} = a \Rightarrow a^{x+1} =?$

A)$\dfrac{1}{6}$ B)3   C)$\dfrac{5a}{3}$

   D)$\dfrac{a}{3}$ E)$\dfrac{a}{6}$

7. $\dfrac{1}{\sqrt{2}} + \dfrac{\sqrt{5+2\sqrt{3+\sqrt{9}}}}{\sqrt{6}+2} =?$

A)$3\sqrt{2}$ B)3   C) $2\sqrt{2}$

   D)2   E)$\sqrt{2}$

8. $X = \frac{1}{\sqrt{3}+\sqrt{2}} \Rightarrow 5 - 2\sqrt{6} = ?$

A)10X

B)$10X^2$

C)X^2

D) $2X^2$

E) $3X^2$

9. $a=1-\sqrt{3} \Rightarrow \sqrt{a^2} - \sqrt{(b-a)^2} + \sqrt[3]{-a^3} = ?$

   $b=\sqrt{2} - 1$

A)-2

B)-1

C)$\sqrt{3} - \sqrt{2}$

D)$2-\sqrt{3}$

E)$\sqrt{2} + \sqrt{3}$

10. $\frac{\sqrt{144}}{0.6} + \frac{\sqrt{2.56}}{0.2} - \frac{\sqrt{0.64}}{0.4} = ?$

A)2

B)4

C)6

D)8

E)10

11. $\sqrt[3]{a\sqrt{b}} = \sqrt[6]{432} \Rightarrow a = b = ?$

A)5   B)7   C)9

D)12   E)15

$12.\sqrt{2} = 1.41 \Rightarrow \sqrt{18} + \sqrt{27} =?$

$\sqrt{3} = 1.73$

A)9.42   B)9.48   C)10.12

D)10.32   E)10.56

First order equations

ax+b=0

$$\Rightarrow ax = -b$$

$$\Rightarrow x = -\frac{b}{a}$$

Definition

: Let a$\neq o$ and $ax + b = 0,$ $such$ $expressions$

are called first order equations

Solution set: S.S$=\left\{-\dfrac{b}{a}\right\}$

(Example):

$$\frac{x}{2} - 1 = \frac{x}{4} - 2 \quad \Rightarrow (SS) = ?$$

A){-4}   B){-3}   C){-2}   D){-1}
E){0}

(Solution):

$$\frac{x}{2} - 1 = \frac{x}{4} - 2$$

$$\frac{x}{2} - \frac{x}{4} = 1 - 2$$

$$\frac{2x}{4} - \frac{x}{4} = -1$$

$$\frac{x}{4} = -1$$

X=-4$\Rightarrow C = \{-2\}$

Answer A

(Example):

$$\frac{3}{\frac{x}{2}+1} - \frac{4}{\frac{x}{2}-1} = 0 \Rightarrow (SS) =?$$

A)$\{-16\}$ B)$\{-14\}$ C)$\{-8\}$ D)$\{4\}$
E)$\{12\}$

(Solution):

$$\frac{3}{\frac{x}{2}+1} - \frac{4}{\frac{x}{2}-1} = 0$$

$$\frac{3}{\frac{x}{2}+1} = \frac{4}{\frac{x}{2}-1}$$

$$3(\frac{x}{2} - 1) = 4(\frac{x}{2} + 1)$$

$$\frac{3x}{2} - 3 = \frac{4x}{2} + 4$$

$$\frac{3x}{2} - \frac{4x}{2} = +3 + 4$$

$$\frac{-x}{2} = 7$$

-x=14

x=-14$\Rightarrow (SS) = \{-14\}$

Answer B

(Example):

$$\frac{x}{\frac{1}{2}+1} + 1 = \frac{3x+1}{3} \Rightarrow (SS) =?$$

A){-2}   B){-1}   C){0}   D){1}

E){2}

(Solution):

$$\frac{x}{\frac{1}{2}+1} + 1 = \frac{3x+1}{3}$$

$$\frac{x}{\frac{3}{2}} + 1 = \frac{3x}{3} + \frac{1}{3}$$

$$\frac{2x}{3} - \frac{3x}{3} = -1 + \frac{1}{3}$$

$$\frac{-x}{3} = \frac{-2}{3}$$

X=2 $\Rightarrow$ $(SS) = \{2\}$

-Answer E

$$\left.\begin{array}{l} a_1 x + b_1 y = c_1 \\ a_2 x + b_2 y = c_2 \end{array}\right\} \text{ These are called equation systems.}$$

(Example):

$$\left.\begin{array}{l} \frac{x+y-5}{5} = \frac{3}{2} \\ \frac{x-y+4}{4} = \frac{7}{5} \end{array}\right\} \Rightarrow x =?$$

A)$\frac{125}{17}$ B)$\frac{123}{23}$ C)7   D)$\frac{141}{20}$

E)9

(Solution):

$$\frac{x+y-5}{5} = \frac{3}{2} \Rightarrow 2x + 2y - 10 = 15 \Rightarrow 5/2x + 2y = 25$$

$$\frac{x-y+4}{4} = \frac{7}{5} \Rightarrow 5x - 5y + 20 = 28 \Rightarrow 2/5x - 5y = 8$$

....

10x+10y=125

+ 10x-10y=16

....

20x=141

$X=\frac{141}{20}$

-Answer D

(Example):

$$\left.\begin{array}{l} 2x + y = 1 \\ 5x - 6y = -28 \end{array}\right\} \Rightarrow (SS) =?$$

A){(-2,3)}   B){(-1,4)}   C){(2,3)}

D({(-2,4)}   E){(1,-3)}

(Solution):

2x+y=-1

Y=-1-2x

5x-6(-1-2x)=-28

5x+6+12x=-28

79

$$17x=-34$$

$$X=-2$$

$$2.(-2)+y=-1$$

$$Y=3$$

$$(SS)=\{(-2,3)\}$$

-Answer A

(Example):

$$\left.\begin{array}{l} x - \frac{y}{3} = 10 \\ \frac{x}{3} + \frac{y}{4} = -1 \end{array}\right\} \Rightarrow (SS) = ?$$

A){(-1,6)}   B){(-2,4)}   C){(2,3)}   D){(6,-12)}   E){(-6,8)}

(Solution):

$$x - \frac{y}{3} = 10 \Rightarrow x = 10 + \frac{y}{3}$$

$$\frac{1}{3} \cdot x + \frac{y}{4} = -1 \Rightarrow \frac{1}{3} \cdot \left(10 + \frac{y}{3}\right) + \frac{y}{4} = -1$$

$$\frac{10}{3} + \frac{y}{9} + \frac{y}{4} = -1$$

$$\frac{13}{36} = -1 - \frac{10}{3}$$

$$\frac{13y}{36} = \frac{-13}{3}$$

$$Y = \frac{-1}{3} \cdot \frac{36}{13}$$

$$Y = -12$$

$$x - \frac{(-2)}{3} = 10$$

X=6

(SS)={(6,12)}

PROPERTIES OF SOLUTION SET

1. $\frac{a_1}{a_2} \neq \frac{b_1}{b_2} \neq \frac{c_1}{c_2}$, *then the solution set has one member.*

(Example):

$$3x + 5y = 29 \atop 2x - 3y = -6 \Bigg\} \Rightarrow (SS) = ?$$

A){(1,2)}   B){(2,3)}   C){(3,4)}
D){(2,4)}   E({{(1,3)}

(Solution):

$3/3x + 5y = 29$

$5/2x - 3y = -6$

....

$\quad 9x + 15y = 87$

+10x-15y=30

....

$\quad$ 19x=57

$\quad$ X=3

$\Rightarrow 2.3 - 3y = -6$

$\qquad$ -3y=-12

$\qquad$ Y=4

(SS)={(3,4)}

$\qquad$ -Answer C

2. $\frac{a_1}{a_2} = \frac{b_1}{b_2} \neq \frac{c_1}{c_2}$ then the solution set is an empty set.

(Example):

$$\left.\begin{array}{l} 2x - 3y = 6 \\ -x + \frac{3}{2}y = -12 \end{array}\right\} \Rightarrow (SS) =?$$

A){(1,2)}   B){(-1,2)}   C){(2,5)}   D)R

E)∅

(Solution):

$$\frac{2}{-1} = \frac{-3}{\frac{3}{2}} \neq \frac{6}{-12} \Rightarrow (SS) = \emptyset$$

-Answer E

3. $\frac{a_1}{a_2} = \frac{b_1}{b_2} =$

$\frac{c_1}{c_2}$ then the solution set has infinite number of members

(Example):

$$\left.\begin{array}{l} 6x - 3y = -6 \\ 4x - 2y = -4 \end{array}\right\} \Rightarrow (SS) =?$$

A){(-2,2)}   B){(-3,3)}   C){(1,5)}   D)R

E)∅

(Solution):

$$\frac{6}{4} = \frac{-3}{-2} = \frac{-6}{-4} \Rightarrow$$

Solution set has infinite number of elements.

TEST WITH SOLUTIONS

1. $x - (2 - x).(4 + x) = -(2 - x).(2 + x) \Rightarrow x =?$

A)-4   B)-3   C)$\frac{1}{4}$ D)$\frac{4}{3}$ E)2

(Solution):

x-(2-x).(4+x)=-(2-x).(2+x)

x-(8+2x-4x+x^2) $= -(4 + 2x - 2x - x^2)$

$x - 8 + 2x + x^2 = -4 + x^2$

3x-8=-4

3x=4

X=$\frac{4}{3}$

-Answer D

2. $3.(x - 1) = (x - 3) + 2x \Rightarrow (SS) =?$

A)∅   B){0}   C){1}   D){2}   E)R

(Solution):

3.(x-1)=(x-3)+2x

3x-3=x-3+2x

3x-3=3x-3

0=0

(SS)=R

-Answer E

3.3ax-b=bx-3a$\Rightarrow x =$?

A)-a   B)-1   C)1   D)$\frac{3}{2}$ E)a

(Solution):

3ax-b=bx-3a

3ax-bx=b-3a $\Rightarrow x = \frac{(3a-b)}{3a-b}$

$= \frac{b-3a}{3a-b}$

$= \frac{-(3a-b)}{3a-b}$

$=-1$

-Answer B

14. $\left.\begin{array}{l} 5^b : 3^a = 30 \\ 3^a : 4 = \frac{1}{24} \end{array}\right\} \Rightarrow b =$?

A)1   B)2   C)3   D)4   E)5

(Solution):

$\frac{3^a}{4} = \frac{1}{24} \Rightarrow 3^a = \frac{1}{6}$

$5^b : \frac{1}{6} = 30 \Rightarrow 5^b . 6 = 30$

$5^b = 5 \Rightarrow b = 1$

-Answer A

15.y-2$[y - 2.[y - 2.(y - 2)]$=21$\Rightarrow y =$?

A)-2   B)-1   C)0   D)1   E)2

86

(Solution):

$$y\text{-}2\left[y - 2[y - 2.(y - 2)] = 21\right.$$

y-2{y-2.[$y - 2y + 4$]} = 21

y-2.(y-2(-y+4))=21

y-2.(y+2y-8)=21

y-2.(3y-8)=21

-5y=21-16

-5y=5$\Rightarrow y = -1$

-Answer B

16. $\dfrac{a-2}{3} - 0.4.\dfrac{(a-1)}{3} = \dfrac{2}{3} \Rightarrow a =$?

A)6   B)7   C)8   D)26   E)32

(Solution): $\dfrac{a-2}{3} - \dfrac{4}{10}.\dfrac{a-1}{3} = \dfrac{2}{3}$

$\dfrac{5a-10-2(a-1)}{15} = \dfrac{2}{3}$

$\dfrac{5a-10-2a+}{15} = \dfrac{2}{3}$

$\dfrac{3a-8}{15} = \dfrac{2}{3}$

9a-24=30

9a=54$\Rightarrow a = 6$

-Answer A

87

17. $\left.\begin{array}{l} a - b = 2 \\ c - d = 4 \\ a + c = 10 \end{array}\right\} \Rightarrow a + b + c + d =?$

A)10 B)11 C)12 D)13 E)14

(Solution):

$\left.\begin{array}{l} a - b = 2 \\ c - d = 4 \\ a + c = 10 \end{array}\right\}$

$a + c = 10$ $\qquad$ a+c=10

$-1/a\ a - b = 2$ $\qquad$ $-1/c - d = 4$

....

a+c=10 $\qquad$ a+c=10

-a+b=-2 $\qquad$ -c+d=-4

....

b+c=8 $\qquad$ a+d=6

a+b+c+d=6+8

=14

Yanit-

Answer E

18. $\left.\begin{array}{l} 2x = 3y \\ \frac{x}{3} + 2y = 10 \end{array}\right\} \Rightarrow y =?$

A)1 B)2 C)4 D)6 E)8

(Solution):

2x=3y

2x-3y=0

$\frac{x}{3} + 2y = 10$

$\frac{x+6y}{3} = \frac{10}{1}$

X+6y=30

$$\begin{array}{c} \frac{2/2x-3y=0}{x+6y=3} \\ -4/\underline{4x-6y=0} \\ 5x=30 \end{array}$$

X=6$\Rightarrow 206 = 3y \Rightarrow y = 4$

-Answer C

19. $\frac{x-1}{2} + \frac{x+1}{4} = 2x - \frac{x-1}{4} \Rightarrow x =$?

A)$-\frac{1}{2}$ 　　　　 B) $-\frac{1}{3}$ 　　　　 C)$\frac{1}{4}$ 　　　　 D)$\frac{1}{3}$

E)$\frac{1}{2}$

(Solution):

$$\underset{(2)}{\frac{x-1}{2}} + \underset{(1)}{\frac{x+1}{4}} = \underset{(4)}{\frac{2x}{1}} - \underset{(1)}{\frac{x-1}{4}}$$

$$\frac{2x-2+x+1}{4} = \frac{8x-x+1}{4}$$

$$\frac{3x-1}{4} = \frac{7x+1}{4}$$

$$3x - 7x = 1 + 1$$

-4x=2

$$x = -\frac{2}{4}$$

$$x = -\frac{1}{2}$$

-Answer A

20. $\left.\begin{array}{l} 3x - 5y + 2z = -11 \\ 6x - 2y + 5z = 7 \end{array}\right\} \Rightarrow x + y + z =?$

A)1   B)2   C)3   D)4   E)6

(Solution):

$-1/3x - 5y + 2z = -11$

$\dfrac{6x-2y+5z=7}{-3x+5y-2z=11}$

$\dfrac{+6x-2y+5z=7}{3x+3y+3z=1}$

3.(x+y+z)=18

X+y+z=6

Answer E

21. $\left.\begin{array}{l} a + b = 14 \\ b + c = 12 \\ a + c = 16 \end{array}\right\} \Rightarrow 2a + b - c =?$

A)12   B)13   C)14   D)15   E)16

(Solution):

a+b=14

b+c=12

$$\frac{-1/a+c=16}{2b=10}$$

b=5$\Rightarrow a = 9 \Rightarrow c = 7$

2.(9)+5-7=16

-Answer E

QUESTIONS

1. $f(x) = \dfrac{2}{x-2} - \dfrac{1}{3}$, $\quad f(x_1) = 0 \Rightarrow x_1 = ?$

A)8   B)6   C)4   D)3   E)1

(Solution):

$f(x) = \dfrac{2}{x-2} - \dfrac{1}{3}$

$f(x_1) = \dfrac{2}{x_1-2} - \dfrac{1}{3} = 0, \dfrac{2}{x_1-2} = \dfrac{1}{3} \Rightarrow x_1 - 2 = 6$

$\qquad x_1 = 8$

2. $\left. \begin{array}{l} a + b = 2 \\ b + c = \dfrac{5}{4} \\ a + c = \dfrac{9}{4} \end{array} \right\} \Rightarrow \dfrac{c}{a} = ?$

A)$\dfrac{3}{2}$ B)$\dfrac{4}{3}$ C)$\dfrac{5}{2}$ D)$\dfrac{2}{3}$

E)$\dfrac{1}{2}$

(Solution):

$a+b=2$

$b+c=\dfrac{5}{4}$

$a+c=\dfrac{9}{4}$

$+$

....

$2a+2b+2c = \underset{(4)}{\dfrac{2}{1}} + \underset{(1)}{\dfrac{5}{4}} + \underset{(1)}{\dfrac{9}{4}}$

92

2.$(a+b+c)=\dfrac{22}{4}$

$\dfrac{a+b+c}{2} = \dfrac{11}{4}$

$2+c=\dfrac{11}{4} \Rightarrow c = \dfrac{11}{4} - 2 = \dfrac{3}{4}$

$a+c=\dfrac{9}{4}$

$a+\dfrac{3}{4} = \dfrac{9}{4} \Rightarrow a = \dfrac{6}{4}$

$\dfrac{c}{a} = \dfrac{\frac{3}{4}}{\frac{6}{4}} \quad \Rightarrow \dfrac{c}{a} = \dfrac{3}{4}\cdot\dfrac{4}{6}$

$\dfrac{c}{a} = \dfrac{1}{2}$

3. $f(x)=x^2 - ax + a - b, f(2) = 0, f(3) = 2 \Rightarrow b =?$

A)-2   B)-1   C)0   D)1   E)2

(Solution):

$f(x)=x^2 - ax + a - b$

$f(2)=2^2 - 2a + a - b$

$f(2)=4-a-b \Rightarrow 4 - a - b = 0 \Rightarrow a + b = 4$

$f(3)=3^2 - 3a + a - b = 2 \Rightarrow 9 - 2a - b = 2$

$\Rightarrow 2a + b = 7$

$-a - b = 4$

$\dfrac{+2a+b=7}{a=3}$

93

b=1

4. $\left.\begin{array}{l}5 - x = y \\ 2x - 6y = 2\end{array}\right\} \Rightarrow (x, y) =?$

A){-3,8}   B){-3,1}   C) (4,1)   D) (4,8)
E) (9,-4)

(Solution):

5-x=y⇒ $x + y = 5$

2x-6y=2⇒ $\dfrac{-1/x-3y=1}{x+y=5}$

$\dfrac{+ \quad\quad -x+3y=-1}{4y=4}$

Y=1

X=4

(1,4)

5. $\dfrac{3}{b} = \dfrac{5}{d} . \dfrac{3}{4} . d = \dfrac{5}{4} . b + c - 1 \Rightarrow c =?$

A)0   B)1   C) 2   D)3   E)4

(Solution):

$\dfrac{3}{b} = \dfrac{5}{d} \Rightarrow 3d = 5b \Rightarrow 3d - 5b = 0$

$\dfrac{3}{4}d = \dfrac{5}{4}b + c - 1$

$\dfrac{3d}{4} - \dfrac{5b}{4} = c - 1$

94

$$\frac{3d-5b}{4} = c - 1$$

$$\frac{0}{4} = c - 1$$

0=c-1

C=1

$$6. \left.\begin{array}{r} 2x + y = z \\ x + y + z = 12 \\ x + z = 3y \end{array}\right\} \Rightarrow z = ?$$

A)3　　　　B)4　　　　　C)5　　　　D)6　　　　E)7

(Solution):

X+z=3y

X+z+y=12

3y+y=12

　Y=3

2x+y=z　⇒　　$2x - z = -3$

2x+3=z　　　$\dfrac{x+z=9}{3x=6}$

　　　　3x=6

　　　　　X=2,　z=7

$$7.\left.\begin{array}{r} K + 2L + M = 6 \\ 2K - L + 2M = 7 \end{array}\right\} \Rightarrow K + L + M = ?$$

A)1　　　　　B)2　　　　　C)3　　　　D)4　　　　E)5

(Solution):

95

$$-2\Big/ \begin{matrix} K + 2L + M = 6 \\ 2K - L + 2M = 7 \end{matrix} \Big\} \Rightarrow -2K - 4L - 2M = -12$$

$$+ \quad \underline{2K - L + 2M = 7}$$
$$-5L = -5$$
$$L = 1$$

K+2.1+M=6$\Rightarrow$ $K + M = 4$

K+L+M=5

8.f(x)=ax+b,f(1)=-2,f(2)=1 $\Rightarrow f(3) =?$

A)1   B)2   C)3   D)4   E)5

(Solution):

f(x)=ax+b

f(1)=a+b=-2$\Rightarrow -1/a + b = -2$

f(2)=2a+b=1$\Rightarrow \dfrac{\begin{matrix}2a+b=1\\ -a-b=2\end{matrix}}{\dfrac{2a+b=1}{a=3}}$

a=3

b=-5

f(3)=3.3-5

f(3)=4

9.x> 0,$\dfrac{x^{\frac{1}{2}}}{3} + \dfrac{x^{\frac{1}{3}}}{x} = \dfrac{x}{6} \Rightarrow x =?$

A)1   B)2   C)3   D)4   E)5

(Solution):

$$\frac{\frac{1}{x}}{3} + \frac{\frac{1}{3}}{x} = \frac{x}{6}$$

$$3x^2 = 12 \Rightarrow x^2 = 4 \Rightarrow x = \pm 2, x = 2$$

10. $\dfrac{2}{\frac{1}{x}} - \dfrac{1}{\frac{2}{x}} = 6 \Rightarrow x = ?$

A)6 B)5 C)4 D)3 E)2

(Solution):

$$\frac{2}{\frac{1}{x}} - \frac{1}{\frac{2}{x}} = 6$$

$$\underset{(2)}{\frac{2x}{1}} - \underset{(1)}{\frac{x}{2}} = 6 \Rightarrow \frac{3x}{2} = 6$$

3x=12

X=4

11. $\dfrac{x}{2} - \dfrac{x-1}{4} = 1 \Rightarrow x = ?$

A)1 B)2 C)3 D)4 E)5

(Solution):

$$\underset{(2)}{\frac{x}{2}} - \frac{x-1}{4} = 1$$

$$\frac{2x-x+1}{4} = 1 \Rightarrow \frac{x+1}{4} = 1$$

X+1=4

X=3

12. $0 < a, 0 < b, 0 < c, \quad \dfrac{b}{a} = \dfrac{1}{3}, \dfrac{a}{c} = \dfrac{2}{3}$

97

a+b+c=34 $\Rightarrow a = ?$

A)8   B)10   C)12   D)14   E)16

(Solution):

$$\frac{b}{a} = \frac{1}{3} \Rightarrow b = \frac{a}{3}$$

$$\frac{a}{c} = \frac{2}{3} \Rightarrow c = \frac{3a}{2}$$

a+b+c=34

$$\underset{(6)}{\frac{a}{1}} + \underset{(2)}{\frac{a}{3}} + \underset{(3)}{\frac{3a}{2}} = 34$$

$$\frac{6a}{6} + \frac{2a}{6} + \frac{9a}{6} = 34$$

$$\frac{17a}{6} = \frac{34}{1}$$

17.a=34.6

a=12

13.K> 0, $x = 2K, y = 3K, z = 4K,$

X+y+z=360$\Rightarrow z = ?$

A)180   B)160   C)120   D)80   E)60

(Solution):

X+y+z=360

2K+3K+4K=360

$$\frac{9k}{9} = \frac{360}{9}$$

K=40,   z=4.K,z=4.40=160

14. $\begin{array}{r} x - 0.2y = 0.2 \\ 2x - y = -20 \end{array}\Big\} \Rightarrow x + y =?$

A)21   B)29   C)41   D)48   E)51

(Solution):

$-5/x - 0.2y = 0.2$

$\dfrac{2x-y=-20}{-5x+y=-1}$

$\dfrac{2x-y=-20}{-3x=-2}$

X=7

Y=34,x+y=7+34

$\qquad$ =41

15. $\begin{array}{r} 2a + b = 10 \\ a + 2c = 14 \\ b + c = 7 \end{array}\Big\} \Rightarrow ? <? <?$

A)b$< a < c$ B) b$< c < a$ C) a$< b < c$

D) a$< c < b$ D) c$< a < b$

(Solution):

2a+b=10

$\dfrac{-b-c=-7}{2a-c=3}$

a+2c=14

$\dfrac{4a-2c=6}{5a=20}$

99

a=4 ,b=2,c=5, b< a < c

16. $0.003 = \frac{1}{100} \cdot k \Rightarrow k =?$

A)$\frac{3}{10}$ B)$\frac{3}{100}$ C)$\frac{3}{20}$ D)$\frac{1}{30}$ E)30

(Solution):

K=0.003.100

K=0.3=$\frac{3}{10}$

17. $(3x-1)^2 = 9x^2 + 13 \Rightarrow x =?$

A)3 B)2 C)-3 D)-2 E)$-\frac{1}{2}$

(Solution):

$9x^2 - 6x + 1 = 9x^2 + 13$

-6x+1=13

-6x=12

X=-2

18. $2^{x+4} + 2^{x+1} + 2^x = 304 \Rightarrow x =?$

A)3 B)4 C)5 D)6 E)7

(Soution):

$2^{x+4} + 2^{x+1} + 2^x = 304$

$2^x \cdot 2^4 + 2^x \cdot 2 + 2^x = 304$

$2^x(16 + 2 + 1) = 304$

$2^x . 19 = 304$

$2^x = 16$

$2^x = 2^4$

X=4

$19.5 \dfrac{5}{x-5} = 10 \Rightarrow x = ?$

A)1   B)2   C)3   D)4   E)6

(Solution):

$5 \dfrac{5}{x-5} = 10$

$\dfrac{5}{\underset{(x-5)}{1}} - \dfrac{5}{\underset{(1)}{x-5}} = 10$

$\dfrac{5x - 25 -}{x-5} = 10$

$\dfrac{5x - 10}{x-5} = 10$

5x-30=10x-50

20=5x

4=x

$1. \dfrac{3x}{2} - \dfrac{2x}{5} = 6\left(\dfrac{4x}{5} - 1\right) \Rightarrow (SS) = ?$

A)$\{\dfrac{1}{2}\}$ B)$\{\dfrac{48}{11}\}$ C)$\{\dfrac{37}{60}\}$ D)$\emptyset$ E)$\{\dfrac{60}{37}\}$

$2.\left(2 - \dfrac{a+1}{a-1}\right):2 = \dfrac{1}{3} \Rightarrow a = ?$

A)$\dfrac{3}{2}$ B)4   C)5   D)6   E)7

$3.\left.\begin{array}{l} 3a + 4b = 5c \\ 4a + 3b = 4c \\ a + b = 27 \end{array}\right\} \Rightarrow c = ?$

A)91   B)24   C)23   D)22   E)21

$4.\left.\begin{array}{l} 2a + b = 16 \\ a + c = 6 \\ b - c = 8 \end{array}\right\} \Rightarrow 4a + 4b - 2c = ?$

A)48   B)49   C)51   D)54   E)56

$5.\dfrac{1}{1+\dfrac{1}{\dfrac{3}{1+\dfrac{2}{x}}}} = \dfrac{3}{2} \Rightarrow x = ?$

A)-4   B)-2   C)0   D)3   E)4

$6.\left.\begin{array}{l} x + y = 3 \\ y + z = 4 \\ z + x = 7 \end{array}\right\} \Rightarrow x + y + z = ?$

A)1   B)3   C)4   D)6   E)7

7. $\dfrac{1-\dfrac{x}{1+\frac{3}{3}}}{3} = \dfrac{1}{3} \Rightarrow x = ?$

A)1   B)2   C)3   D)4   E)6

8. $x - \dfrac{2}{x} : \dfrac{4}{3} = 4x \Rightarrow x = ?$

A)$\dfrac{1}{2}$ B)$-\dfrac{1}{3}$ C)$\dfrac{1}{3}$ D)$-\dfrac{1}{6}$ E) 6

9. $\dfrac{1-3x}{2} - \dfrac{x-2}{3} = 1 \Rightarrow x = ?$

A) $\dfrac{1}{2}$ B) $\dfrac{1}{11}$ C) $\dfrac{11}{7}$ D)18   E)21

10. $2x + \dfrac{1}{3}(x-3) = 6 \Rightarrow x = ?$

A)3   B)10   C)18   D)21   E) $\dfrac{6}{7}$

11. $a \neq b$

$3a + \dfrac{2}{a} = 3b + \dfrac{2}{b} \Rightarrow a.b = ?$

A)2   B)3   C)$\dfrac{2}{3}$ D)$\dfrac{4}{9}$ E)12

12. $1 - \dfrac{1}{x - \frac{2}{3}} = 2 \Rightarrow x = ?$

A)$-\dfrac{1}{2}$ B)$-\dfrac{1}{3}$ C)$\dfrac{1}{3}$ D)6   E)9

13.$\dfrac{b(x-a)}{2} - \dfrac{a(x-2b)}{4} = \dfrac{ab}{4} \Rightarrow x =?$

A)$\dfrac{2b-2a}{ab}$ B)$\dfrac{ab}{2b-a}$ C)$\dfrac{ab-1}{a+b}$ D)$\dfrac{2b}{a-b}$ E)$\dfrac{a+b}{a-b}$

14. $\left.\begin{array}{l} 2a - b + 3c = 6 \\ a + 2b + c = 4 \\ 2a + 4b + c = 10 \end{array}\right\} \Rightarrow a + b + c =?$

A)1   B)4   C)16   D)20   E)24

15.$\dfrac{2x-2}{3} = \dfrac{b+x}{4} \Rightarrow x =?$

A)$\dfrac{2b+4}{3}$ B)$\dfrac{3b+8}{5}$ C)$\dfrac{11b-8}{3}$ D)$\dfrac{8-11b}{5}$

E)$\dfrac{5b-8}{3}$

16.$\dfrac{2011.x-2030}{2030-2011.x} = \dfrac{x-7}{2x+1} \Rightarrow x =?$

A)0   B)1   C)2   D)2000   E)2019

17.$\dfrac{x}{a} + \dfrac{x}{b} = \dfrac{b}{x} + \dfrac{a}{x} \Rightarrow x =?$

A)$\dfrac{a}{b}$ B)a^2b C)ab   D)$\sqrt{ab}$ E)$\sqrt{\dfrac{a}{b}}$

18. $\left(2 - \dfrac{a+3}{a-2}\right) = 3 \Rightarrow a =?$

A)$-\dfrac{1}{2}$ B)$\dfrac{3}{4}$ C)5 D)6 E)$\dfrac{7}{3}$

19. $a \neq 0$

$a(4-2x)=-2(ab+ax) \Rightarrow b =?$

A)0 B)1 C)-2 D)4 E)8

20. $\left.\begin{array}{r} 2ax + 2by = 17 \\ 4ax - 2by = 19 \\ x = y = 3 \end{array}\right\} \Rightarrow a - b =?$

A)$\dfrac{7}{6}$ B)$\dfrac{-4}{3}$ C)$\dfrac{10}{3}$ D)8 E)18

21. $\left.\begin{array}{r} a + b - c + d = 3 \\ 2a + 3c + 2d + b = 6 \\ 3a + 4b + 4c + 3d = 9 \end{array}\right\} \Rightarrow a + b + c + d =?$

A)2 B)3 C)12 D)$\dfrac{9}{2}$ E)4

22.

+	a	b	c
a		9	

b   10

c   11

$$\Rightarrow a - b + c = ?$$

A)7   B)9   C)10   D)11   E)12

$$23. \left. \begin{array}{l} x^2 . y = 6 \\ y^2 . z = 12 \\ z^2 . x = 3 \end{array} \right\} \Rightarrow x . y . z = ?$$

A)3   B)4   C)6   D)36   E)21

(Answers)

1.E	2.E	3.E	4.A	5.A	6.E
7.C	8.D	9.B	10.E	11.C	12.B
13.B	14.B	15.B	16.C	17.D	18.A
19.C	20.A	21.B	22.A	23.C	

1. $\dfrac{a-\frac{1}{a}}{2-\frac{2}{a}} = \dfrac{2a-4}{4} \Rightarrow (SS) =?$

A){4}　　　B){2}　　　C){-2}　　　D){-4}　　　E)∅

2. $\dfrac{a+2}{a-1} - \dfrac{2-a}{1-a} = 1 \Rightarrow a =?$

A)5　　　B)3　　　C)2　　　D)1　　　E)0

3. $\dfrac{3x+2}{4} - \dfrac{4x-3}{4} = 2 \Rightarrow x =?$

A)-3　　　B)$\dfrac{1}{3}$　　　C)3　　　D)1　　　E)$\dfrac{1}{2}$

4. (x+a)(x-a)=$2x^2 - (x-a)^2, a \neq 0 \Rightarrow x =?$

A)a　　　B)-a　　　C)0　　　D)$\dfrac{a}{2}$　　　E)$-\dfrac{a}{3}$

5. $\dfrac{x-1}{2} + \dfrac{x+1}{4} = 2x - \dfrac{x-1}{4} \Rightarrow x =?$

A)$\dfrac{1}{3}$　　　B)$\dfrac{1}{2}$　　　C)$-\dfrac{1}{3}$　　　D)$-\dfrac{1}{2}$　　　E)$\dfrac{1}{4}$

6. $\dfrac{2}{2+\frac{2}{2+\frac{2}{x}}} = 2 \Rightarrow x =?$

A)0　　　B)1　　　C)$-\dfrac{1}{2}$　　　D)-2　　　E)$\dfrac{1}{2}$

7.12-2a-6$\sqrt{2}$ $+ \frac{a}{6} + 2a - 12 + 5\sqrt{2} = 0 \Rightarrow a = ?$

A)$6\sqrt{2}$ B)2   C)0   D)12   E)8

8.0.2+3=x$+\frac{1}{3} \Rightarrow x = ?$

A)-2   B)0   C)2   D)$\frac{24}{7}$ E)$\frac{13}{3}$

9.$\frac{0}{x-1} - \frac{1}{x+1} = \frac{1}{x} - \frac{1}{x+2} \Rightarrow x = ?$

A)1   B)$\frac{1}{2}$ C)0   D)-1   E)$-\frac{1}{2}$

10.$\frac{0.2x-0.1}{x-0.8} = 0.1 \Rightarrow x = ?$

A)$-\frac{5}{3}$ B)$-\frac{3}{5}$ C)0   D)$\frac{1}{5}$ E)$\frac{3}{5}$

11.$\frac{1}{a} + \frac{1}{x} = \frac{1}{b} \Rightarrow x = ?$

A)0   B)ab   C)$\frac{1}{ab}$ D)$\frac{ab}{a-b}$ E)a+b

12.$(2 - x)^2 = x(x - 2) \Rightarrow x = ?$

A)-2   B)0   C)2   D)4   E)5

$13.\ 2^{x-1}.5^{x+1} = \left(\frac{1}{5}\right)^{-2} \Rightarrow x = ?$

A)-3 B)-1 C)0 D)1 E)2

$14.\ 2^x + 3.2^x + 2^{x+2} = 64 \Rightarrow x = ?$

A)1 B)2 C)3 D)4 E)$\frac{1}{2}$

$15.\ 3^{x-2} + \sqrt{9^{x-2}} = \frac{2}{27} \Rightarrow x = ?$

A)-2 B)-1 C)2 D)3 E)$\frac{1}{3}$

$16.\ \frac{4.8}{0.x} + \frac{1.2}{20} = 4.08 \Rightarrow x = ?$

A)16 B)14 C)12 D)8 E)4

$17.\ 1 + \cfrac{1}{1-\frac{x}{x+1}} = 4 \Rightarrow x = ?$

A)2 B)3 C)4 D)5 E)6

$18.\ 3^x \sqrt[3]{3} = 9\sqrt{3} \Rightarrow x = ?$

A)$\frac{1}{2}$ B)$\frac{1}{3}$ C)$\frac{1}{9}$ D)$\frac{13}{6}$ E)$\frac{11}{2}$

19. $\frac{\frac{1}{2}-x}{1-x}=0.1 \Rightarrow x=?$

A)$\frac{1}{2}$ B)$\frac{4}{9}$ C)1   D)2   E)$\frac{9}{4}$

20. $-2^{-2}+2^{-1}=2^{-x+2} \Rightarrow x=?$

A)1   B)2   C)3   D)4   E)5

21. $\frac{a+\frac{b}{c}}{c+\frac{b}{a}}=ax \Rightarrow x=?$

A)$\frac{1}{a}$ B)$\frac{1}{b}$ C)$\frac{1}{c}$ D)a   E)c

22. $\frac{a+b}{a-b}=\frac{x}{b-a} \Rightarrow x=?$

A)a+b   B)-1   C)a^2-b^2 D)a-b   E)-a-b

23. $(0.4).x=5 \Rightarrow (1.44).x=?$

A)5.055   B)0.55   C)16   D)32   E)18

24) $\frac{x!}{x^2-3x}=\frac{(x-1)!}{13} \Rightarrow x=?$

A)10   B)13   C)16   D)19   E)22

(Answers)

1.E	2.A	3.A	4.C	5.D	6.C
7.A	8.A	9.E	10.D	11.D	12.C
13.D	14.C	15.B	16.A	17.C	18.D
19.B	20.D	21.C	22.E	23.E	24.C

1. $(1.1)^2 - (0.1)^2 = 0.2. x \Rightarrow x = ?$

A)1　　　B)2　　　C)3　　　D)4　　　E)5

2. $x. \sqrt[3]{0.027} = \frac{1}{3} \Rightarrow x = ?$

A)3　　　B)2　　　C)1　　　D)$\frac{10}{9}$　　　E)$\frac{1}{2}$

3. $12! - x. (10!) = 130. (10!) \Rightarrow x = ?$

A)1　　　B)2　　　C)3　　　D)4　　　E)5

4. $\frac{x}{0.03} = y, 100 < y < 1000 \Rightarrow ? < x < ?$

A)3<x<10　　　　B)3<x<30　　　　C)3<x<100

D)10<x<30　　　　E)10<x<100

5. $\left.\begin{array}{l} \frac{x}{5} = \frac{y}{7} \\ \frac{5y}{3} = \frac{7x}{3} + z - 2 \end{array}\right\} \Rightarrow z = ?$

A)0　　　B)1　　　C)2　　　D)3　　　E)4

6. $\frac{3y-1}{y^2-y} - \frac{2}{y-1} = \frac{3}{y} \Rightarrow (SS) = ?$

A){0}　　　B){1}　　　C){2}　　　D)Ø　　　E){3}

7. $0.42-0.031=3.33-t \Rightarrow t =?$

A)0.3   B)0.7   C)1   D)2   E)3

8. $\dfrac{x}{\frac{2x-3}{4}} - \dfrac{4}{\frac{3-2x}{3}} = 4 \Rightarrow x =?$

A)1   B)2   C)4   D)5   E)6

9. $\dfrac{1+\frac{1}{x-1}}{1-\frac{1}{x}} = 1 \Rightarrow x =?$

A)$\frac{1}{2}$ B)-2   C)$-\frac{1}{3}$ D)$\frac{2}{3}$ E)4

10. $1-\dfrac{1}{1+\frac{1}{x-1}} = 2 \Rightarrow x =?$

A)-2   B)-1   C)$\frac{1}{2}$ D)$\frac{3}{4}$ E)4

11. $\dfrac{x-\frac{x}{0.2}}{0.3} + 36 = 0 \Rightarrow x =?$

A)0.1   B)0.6   C)1   D)2   E)3

12. $\dfrac{1}{x+1} + \dfrac{1}{x-1} = \dfrac{x-3}{x^2-1} \Rightarrow x =?$

A)-3   B)-2   C)0   D)1   E)6

13. $\left(t^2 - \frac{1}{x^2}\right) : \left(t - \frac{1}{x}\right) = 1 \Rightarrow t = ?$

A)$\frac{x-1}{x}$ B)x+1 C)$\frac{x}{x+1}$ D)$\frac{x-1}{x+1}$ D)4x

14. $\frac{8x^2}{x^3-a^3} : \frac{4x^2}{x^2+ax+a^2} = 3 \Rightarrow x = ?$

A)a B)$\frac{a}{2}$ C)a+2 D)2a+2 E)3a

15. $ax+b^2 = a^2 + bx \Rightarrow x = ?$

A)a-b B)a C)$\frac{2b}{a}$ D)$\frac{a}{a-b}$ E)a+b

16. $\left(\frac{1}{x^2} - \frac{1}{a^2}\right) : \left(\frac{1}{x} + \frac{1}{a}\right) = 1 \Rightarrow x = ?$

A)a B)2a C)$\frac{a}{a+1}$ D)$\frac{a-1}{a}$ E)$\frac{a+1}{a-1}$

17. $\frac{y}{2} + \frac{y}{3} + \frac{y}{4} = y - 1 \Rightarrow y = ?$

A)-24 B)-18 C)-15 D)-14 E)-12

18. $\frac{x^2-1}{x^2+2x-3} = 2 \Rightarrow x = ?$

A)-9 B)-5 C)-1 D)4 E)7

19. $4-(1-x)^2 = (3-x)(3-x) \Rightarrow x =?$

A)0   B)1   C)2   D)3   E)6

20. $\frac{x+1}{3} - \frac{3x-1}{5} = x - 2 \Rightarrow x =?$

A)1   B)2   C)3   D)4   E)5

21. $\left(\frac{2}{1-x} + \frac{3}{x+1}\right)^{-1} = -1 \Rightarrow x =?$

A)-4   B)-2   C)-1   D)2   E)3

22. $0.2x+0.02x=0.04x+18 \Rightarrow x =?$

A)1   B)10   C)100   D)110   E)1000

23. $\frac{1-0.2x}{3} = x - 0.3 \Rightarrow x =?$

A)0.2   B)2   C)$\frac{25}{9}$ D)3.2   E)$\frac{18}{29}$

24. $\frac{1-x}{1-\frac{1}{x}} + \frac{x}{1+\frac{1}{x}} - \frac{1}{x+1} = \frac{x-6}{x} \Rightarrow x =?$

A)-2   B)0   C)2   D)3   E)9

(Answers)

1.E	2.D	3.B	4.B	5.C	6.D
7.E	8.E	9.A	10.C	11.E	12.A
13.A	14.E	15.E	16.C	17.E	18.B
19.B	20.B	21.D	22.C	23.E	24.D

1. $\dfrac{0.22}{x} = \dfrac{0.11}{0.12} \Rightarrow x =?$

A)0.24   B)0.22   C)0.18   D)0.12   E)0.11

2. 2-(2-(2-x))=x$\Rightarrow x =?$

A)-2   B)-1   C)0   D)1   E)2

3. $\dfrac{x}{x-3} = \dfrac{x-2}{x-6} \Rightarrow x =?$

A)-2   B)-3   C)-4   D)-5   E)-6

4. $\dfrac{2}{2+\dfrac{2}{2+\dfrac{2}{x}}} = 2 \Rightarrow x =?$

A)$-\dfrac{1}{2}$ B)$-\dfrac{1}{4}$ C)$\dfrac{1}{4}$ D)$\dfrac{1}{2}$ E)1

5. $\left.\begin{array}{l} 2x + 4 = 0 \\ 3x - y + 1 = 0 \end{array}\right\} \Rightarrow y =?$

A)-7   B)-5   C)0   D)1   E)2

6. $\left.\begin{array}{l} \dfrac{2}{a} + \dfrac{3}{b} = 3 \\ \dfrac{4}{a} - \dfrac{3}{b} = 1 \end{array}\right\} \Rightarrow a =?$

A)$\dfrac{2}{3}$ B)1   C)$\dfrac{3}{2}$ D)$\dfrac{5}{2}$ E)4

7. $\left.\begin{array}{l}2x + y - z = 8\\ x + 2y + z = 4\end{array}\right\} \Rightarrow y + z =?$

A)-2   B)0   C)$\frac{2}{5}$ D)$\frac{1}{2}$ E)2

8. $(a - 1)^2 = (a - 2)(a - 4) \Rightarrow a =?$

A)-2   B)$-\frac{7}{4}$ C)$\frac{7}{4}$ D)2   E)$\emptyset$

9. $\left.\begin{array}{l}2x + 3y = 16\\ 2y + c = 8\\ 3x - c = 6\end{array}\right\} \Rightarrow 2x + y =?$

A)10   B)9   C)8   D)5   E)1

10. $\dfrac{12}{x+a} + \dfrac{2}{x-5} - \dfrac{2}{x-6} = \dfrac{2}{3}$ $, x = 8 \Rightarrow a =?$

A)2   B)3   C)4   D)5   E)6

11. $\left.\begin{array}{l}-2x + 3y - 4z = 32\\ 3x - 2y + 5z = 24\end{array}\right\} \Rightarrow x + y + z =?$

A)28   B)36   C)42   D)48   E)56

12. $\left.\begin{array}{l}a + 10b + 14c = 30\\ a + 5b + 7c = 32\end{array}\right\} \Rightarrow a =?$

A)30   B)32   C)34   D)36   E)40

13. $\dfrac{x+y+9}{x+y} = 10 \Rightarrow x =?$

A)1+y   B)1-y   C)$\dfrac{1+y}{2}$ D)$\dfrac{1+y}{9}$ E)$\dfrac{1-y}{9}$

14. $3 - \dfrac{1}{2+\dfrac{x}{1-\frac{1}{2}}} = 2 \Rightarrow x =?$

A)-2   B)$-\dfrac{3}{2}$ C)-1   D)$-\dfrac{1}{2}$ E)1

15. $\left.\begin{array}{l}\dfrac{x+y-4}{4} = \dfrac{5}{4} \\ \dfrac{x-y+3}{3} = \dfrac{4}{3}\end{array}\right\} \Rightarrow x =?$

A)5   B)6   C)7   D)8   E)10

16. $\left.\begin{array}{l}5x - y + z = 5 \\ 2x + y - z = 2\end{array}\right\} \Rightarrow 3x + 2x - 2y =?$

A)2   B)3   C)6   D)10   E)11

17. $a - \dfrac{4a-3}{2} = 4 - \dfrac{3a-2}{4} \Rightarrow a =?$

A)-15   B)-12   C)12   D)15   E)17

18. $\dfrac{1+\frac{x}{x+2}}{1-\frac{x}{x-2}} = 1 \Rightarrow x = ?$

A)-2   B)-1   C)0   D)2   E)3

19. $\left.\begin{array}{l} ax + b = -2 \\ a + bx = 4 \end{array}\right\} \Rightarrow x = ?$

A)5x   B)$\dfrac{11x}{2}$ C)3+4x   D)$\dfrac{3}{x}$ E)$\dfrac{2}{x+1}$

20. $\left.\begin{array}{l} 3x + y + z = 28 \\ x + 3y + z = 25 \\ x + y + 3z = 22 \end{array}\right\} \Rightarrow \dfrac{x+y}{z} = ?$

A)$\dfrac{2}{7}$ B)$\dfrac{7}{23}$ C)$\dfrac{7}{2}$ D)$\dfrac{23}{7}$ E)$\dfrac{15}{7}$

21. $\left.\begin{array}{l} \dfrac{5}{x} - \dfrac{2}{y} = 7 \\ \dfrac{2}{x} - \dfrac{3}{y} = -6 \end{array}\right\} \Rightarrow x + y = ?$

A)$\dfrac{5}{12}$ B)$\dfrac{1}{2}$ C)$\dfrac{7}{12}$ D)$\dfrac{2}{3}$ E)2

22. $\left.\begin{array}{l} 2a + 3b - 2c = 6 \\ a + b = c = 5 \end{array}\right\} \Rightarrow a + 2b - c = ?$

A)-2   B)-1   C)0   D)1   E)2

$$23. \left. \begin{array}{l} 2a + b = 3 \\ b + 3c = 5 \\ c - a = 7 \end{array} \right\} \Rightarrow b = ?$$

A)25 B)29 C)33 D)37 E)41

$$24. \left. \begin{array}{l} a.b = \frac{3}{2} \\ b.c = 12 \\ a.c = 2 \end{array} \right\} \Rightarrow c = ?$$

A)3 B)4 C)5 D)6 E)7

(Answers)

1.A	2.D	3.E	4.A	5.B	6.C
7.B	8.C	9.C	10.C	11.E	12.C
13.B	14.D	15.A	16.A	17.B	18.C
19.E	20.D	21.C	22.D	23.E	24.B

RULES OF INEQUALITIES

1. $\quad a < b \Leftrightarrow a \mp c < \mp c$

2. $\quad a < b \ (and) \ c > 0 \Leftrightarrow a.c < b.c$

3. $\quad a < b \ (and) c < 0 \Leftrightarrow a.c > b.c$

4. $\quad a < b \ (and) c > 0 \Leftrightarrow \frac{a}{c} < \frac{b}{c}$

5. $\quad a < b \ (and) c < 0 \Leftrightarrow \frac{a}{c} > \frac{b}{c}$

6. $\quad a, b \in R^+ \ (and) \ a < b \Leftrightarrow \frac{1}{a} > \frac{1}{b}$

(Example):

$3x - 18 < 0 \Rightarrow (SS) = ?$

(Solution):

3x-18<0

3x<18

X<6

(SS)=$\{x | -\circ < x < 6 (and) x \in Q\}$ (or)

(SS)=$(-\circ, 6)$

(Example):

$\frac{x}{2} - 3 \le \frac{3x-1}{5} \Rightarrow (SS) =?$

(Solution):

$\frac{x}{2} - 3 \le \frac{3x-1}{5}$

$10.\frac{x-6}{2} \le \frac{3x-^+}{5}. 10$

$5(x - 6) \le (3x - 1). 2$

$5x - 30 \le 6x - 2$

$2 - 30 \le 6x - 5x$

$-28 \le x$

(SS)=$(-28, +\infty)$

(Example):

$3 < \frac{5-7x}{6} \leq 5 \quad \Rightarrow (SS) = ?$

(Solution):

$3 < \frac{5-7x}{6} \leq 5$

$6.3 < 6.\frac{5-7x}{6} \leq 6.5$

$18 < 5 - 7x \leq 30$

$-5 + 18 < -5 + 5 - 7x \leq -5 + 30$

$13 < -7x \leq 25$

$\frac{13}{-7} > \frac{-7x}{-7} \geq \frac{25}{-7}$

$\frac{-13}{7} > x \geq -\frac{25}{7}$

$(SS) = \left[-\frac{25}{7}, -\frac{13}{7} \right]$

(Example):

$\left. \begin{array}{l} 3x + 5 \leq 86 \\ \frac{-2x+1}{3} < 7 \end{array} \right\} \Rightarrow (SS) = ?$

(Solution):

124

$$3x + 5 \leq 86$$

$$\frac{-2x+1}{3} < 7$$

$$3x \leq 81$$

$$-2x + 1 < 2^1$$

$$x \leq 27$$

$$-2x < 20$$

$$x > -10$$

(SS)= (-10,27)

(Example):

$$\left.\begin{array}{l} 3x - 7 \geq 2x + 1 \\ 5x + 3 < 3x - 7 \end{array}\right\} \quad \Rightarrow (SS) =?$$

(Solution):

$$3x - 7 \geq 2x + 1$$

$$5x + 3 < 3x - 7$$

$$3x - 2x \geq 7 + 1$$

$$5x - 3x < -3 - 7$$

$$x \geq 8$$

$$2x < -10$$

$$x < -5$$

(SS)=∅

ABSOLUTE VALUE

(Definition): $|x| = \begin{cases} x, & x > 0 \\ 0, & x = 0 \\ -x, & x < 0 \end{cases}$

Such an expression is called absolute value of x and shown by $|x|$.

1. $a \geq 0 \Rightarrow |a| = a$
2. $a < 0 \Rightarrow |a| = -a$

BASIC PROPERTIES

1. $|-5| = 5, |-3,15| = 3,15$
    $|-\sqrt{2}| = \sqrt{2}$

2. $n \in N^+, a \in R - \{0\}$
    $|a^n| = |a|^n$

    If n is an odd natural number, $|a^n| \neq a^n$

    If n is an even natural number, $|a^n| = a^n$

(Example):

1. $|(x-3)^3| \neq (x-3)^3$
2. $|(x-2)^2| = (x-2)^2$
3. $|(-7)^5| \neq (-7)^5$
4. $|(-3)^6| = (-3)^6$

126

3. $a,b \in R^+$, $a > b$

$$\begin{cases} |a - b| = a - b \\ |b - a| = a - b \end{cases}$$

$$|a - b| = |b - a|$$

(Example):

$$b - 3c < 2a\sqrt{2} \ (and) \ |b - 3c - 2a| = 3c - 4a$$

$$\Rightarrow \frac{4a+3b}{b-5a} = ?$$

A)16   B)18   C)20   D)22   E)24

(Solution):

$$b - 3c < 2a$$

$$|b - 3c - 2a| = 3c - 4a$$

$$2a - (b - 3c) = 3c - 4a$$

$$2a - b + 3c = 3c - 5 = 4a$$

$$6 . a = b$$

$$\frac{4a+3b}{b-5a} = \frac{4a+3.6a}{6a-5a} = \frac{22a}{a} = 22$$

127

(Example):

$$3 < x < 8 \Rightarrow |2 - x| + |8 - x| = ?$$

A)2   B)4   C)6   D)8   E)10

(Solution):

$$2 < x \Rightarrow |2 - x| = x - 2$$

$$x < 8 \Rightarrow |8 - x| = 8 - x$$

$$|2 - x| + |8 - x| = x - 2 + 8 - x$$

$$= 6$$

(Example):

$$4 < x < 8 \Rightarrow |3 - x| - |x - 4| - |x - 8| = ?$$

A)x+17   B)x-7   C)3x-7   D)3x+7

E)x+15

$4 < x \Rightarrow |3 - x| = x - 3$

$4 < x \Rightarrow |x - 4| = x - 4$

$x < 8 \Rightarrow |x - 8| = 8 - x$

$|3 - x| - |x - 4| - |x - 8| = x - 3 - (x - 4) - (8 - x)$

$$= x - 3 - x + 4 - 8 + x$$

$$= x - 7$$

(Example):

$4 < x < 9 \Rightarrow |x - |x - 4| + 3| + 2x + 3 = ?$

A)10+2x   B)8+2x   C)6+2x   D)4+2x

E)2+x

(Solution):

$4 < x \Rightarrow |x - 4| = x - 4$

$|x - |x - 4| + 3| + 2x + 3 = |x - (x - 4) + 3| + 2x + 3$

$|x - x + 4 + 3| + 2x + 3$

=10+2x

4.a,b,c∈ R

$|a| + |b| + |c| = 0 \Rightarrow a = 0, b = 0, c = 0$

(Example):

$|x - 3| + |2y + x - 25| = 0 \Rightarrow y =?$

A)7   B)11   C)15   D)14
E)23

(Solution):

$|x - 3| + |2y + x - 25| = 0$

$$|x - 3| = 0 \Rightarrow x = 3$$

$2y + x - 25 = 0 \Rightarrow 2y + 3 - 25 = 0$

$$\Rightarrow 2y = 22$$

$$\Rightarrow y = 11$$

(Example):

$|a - 2| + |b - 4a| + |c - 2b| = 0 \Rightarrow a + 2b + 3c =?$

A)62   B)63   C)64   D)65
E)66

(Solution):

$$a - 2 = 0 \Rightarrow a = 2$$

$$b - 4a = 0 \Rightarrow b = 4.2 = 8$$

$$c - 2b = 0 \Rightarrow c = 2.8 = 16$$

$$a + 2b + 3c = 2 + 2.8 + 3.16$$

$$= 2 + 16 + 48$$

$$=66$$

Absolute value of an expression cannot be equal to a negative number

$$\forall x \in R \Rightarrow \sqrt{x^2} = |x|$$

$$\sqrt{x^4} = x^2, \sqrt{x^8} = x^4, \dots \dots \dots$$

(Example):

$$x < 0, \frac{\sqrt{x^2}}{x} + 3 =?$$

A)1   B)2   C)3   D)4

E)5

(Solution):

$$\frac{\sqrt{x^2}}{x} + 3 = \frac{|x|}{x} + 3$$

$$= \frac{-x}{x} + 3$$

$$-1 + 3$$

$$= 2$$

(Example):

$$x > 3, \frac{\sqrt{x^2 - 6x + 9}}{x^2 - 9} + \frac{1}{x+3}$$

A)x+1   B)x+3   C)$\frac{2}{x+3}$ D)$\frac{x+3}{2}$

E)$\frac{x+3}{4}$

(Solution):

$$\frac{\sqrt{(x-3)^2}}{(x-3).(x+3)} + \frac{1}{x+3} = \frac{|x-3|}{(x-3).(x+3)} + \frac{1}{x+3}$$

$$= \frac{x-3}{(x-3).(x+3)} + \frac{1}{x+3}$$

$$= \frac{2}{x+3}$$

(Example):

$$a, b \in R, a < \frac{b}{2} \Rightarrow \sqrt{(2a-b)^2} - \sqrt{(a-b)^2} = ?$$

A)-a   B)$\frac{a+b}{2}$ C)a+b   D)a+2b

E)2a+2b

(Solution):

$$= \sqrt{(2a-b)^2} - \sqrt{(a-b)^2}$$

$$= |2a-b| - |a-b|$$

$$= b - 2a - (b-a) = -a$$

ABSOLUTE VALUE EQUATIONS

1. $a > 0$

$$|x| = a \Rightarrow \begin{cases} x_1 = a \\ x_2 = -a \end{cases}$$

2. $b \in R^+$

$$|x - a| = b \Rightarrow \begin{cases} x - a = b \Rightarrow x_1 = a + b \\ x + a = -b \Rightarrow x_2 = a - b \end{cases}$$

$$\begin{rcases} x < 0 \\ y > 0 \end{rcases} \Rightarrow |xy| = -xy$$

$$\begin{rcases} x < 0 \\ y < 0 \end{rcases} \Rightarrow |xy| = xy$$

(Example):

$$\left| \frac{2x-1}{5} \right| = 7 \quad \Rightarrow (SS) = ?$$

A){16,17} B){-17,-18} C){-17,18} D){17,18}
E){18,-19}

(Solution):

$$\left|\frac{2x-1}{5}\right| = 7$$

$$\frac{2x-1}{5} = 7 \qquad\qquad\qquad \frac{2x-1}{5} = -7$$

$$2x - 1 = 35 \qquad\qquad\qquad 2x - 1 = -35$$

$$2x = 36 \qquad\qquad\qquad 2x = -34$$

$$x_1 = 18 \qquad\qquad\qquad x_2 = -17$$

$$(SS) = (-17,18)$$

(Example):

$$|x^2 - 1| = 3 \Rightarrow (SS) =?$$

A){1,2}   B){2,2)   C){-2,2)   D){2,3)
E){-2,-5)

(Solution):

$$|x^2 - 1| = 3$$

$$x^2 - 1 = 3 \qquad\qquad\qquad x^2 - 1 = -3$$

$$x^2 = 4 \qquad\qquad\qquad x^2 = -2$$

$$|x| = 2$$

$$\left.\begin{array}{l} x_1 = 2 \\ x_2 = -2 \end{array}\right\} \Rightarrow (SS) = \{-2,2\}$$

(Example):

$$|x - 4| + \sqrt{x^2 - 8x + 16} = 12 \Rightarrow x_1, x_2 = ?$$

A)-25   B)-20   C)-15   D)-10   E)-5

(Solution):

$$|x - 4| + \sqrt{(x - 4)^2} = 2$$

$$|x - 4| + |x - 4| = 2$$

$$2|x - 4| = 12$$

$$|x - 4| = 6$$

$x - 4 = 6$ $\qquad\qquad\qquad$ $x - 4 = -6$

$x_1 = 10$ $\qquad\qquad\qquad$ $x_2 = -20$

$x_1 . x_2 = 10. (-2) = -20$

(Example):

$$A^2 - 7A + 10 = 0 \ (and) \ |x - 3| = A \Rightarrow \sum x = ?$$

A)8   B)9   C)10   D)11   E)12

(Solution):

$$A^2 - 7A + 10 = 0$$

$$(A - 2).(A - 5) = 0$$

$$A = 2 \quad (or) \qquad A = 5$$

$$x - 3 = 2 \rightarrow x_1 = 5$$

$$|x - 3| = 2$$

$$x - 3 = -2 \rightarrow x_2 = 1$$

$$x - 3 = 5 \rightarrow x_3 = 8$$

$$|x - 3| = 5$$

$$x - 3 = -5 \rightarrow x_4 = -2$$

$$x_1 + x_2 + x_3 + x_4 = 5 + 1 + 8 - 2$$

$$= 12$$

ABSOLUTE VALUE INEQUALITIES

1. $a \in R^+$

   $|x| < a \Leftrightarrow -a < x < a$

   $|x| < 5 \Leftrightarrow -5 < x < 5$

   $|x| < \frac{3}{4} \Leftrightarrow -\frac{3}{4} < x < \frac{3}{4}$

   $|x| < 0 \Rightarrow (SS) = \emptyset$

   $|x| < -3 \Rightarrow (SS) = \emptyset$

2. $a \in R^+$

   $|x| > a \Rightarrow \begin{cases} x > a \\ x < -a \end{cases}$

Numbers in the solution set are in the dark region.

(Example):

$|x| < 10 \ (and) x - 2y = 4 \Rightarrow ? < y < ?$

A)-7<y<-3   B)-7<y<-1   C)-7<y<3   D)-6<y<2

E)-6<y<3

(Solution):

$|x| < 10$ $x - 2y = 4$

$-10 < x < 10$ $x = 2y + 4$

$-10 < 2y + 4 < 10$

$-14 < 2y < 6$

$-7 < y < 3$

(Example):

$|x| < 6 \ (and)y = \frac{k}{2} + 4 \Rightarrow \sum y =?$

A)17   B)18   C)19   D)20   E)21

(Solution):

$y = \frac{x}{2} + 4 \Rightarrow y - 4 = \frac{x}{4}$

$\Rightarrow x = 2y - 8$

$|x| < 6 \Leftrightarrow |2y - 8| < 6$

$-6 < 2y - 8 < 6$

$2 < 2y < 14$

$1 < y < 7$

$\sum y = 2 + 3 + 4 + 5 + 6 = 20$

$3.a \in R^+ \ |x - b| < a \Rightarrow -a < x - b < a$

$\Rightarrow b - a < x < a + b$

(Example):

$$\left|\frac{2x-3}{5}\right| > 3 \Rightarrow ? < x < ?$$

A)-6<x<9   B)-6<x<8   C)-7<x<9   D)-8<x<10
E)-8<x<11

(Solution):

$$\left|\frac{2x-3}{5}\right| < 3$$

$$-3 < \frac{2x-3}{5} < 3$$

$$-15 < 2x - 3 < 15$$

$$-12 < 2x < 18$$

$$-6 < x < 9$$

(Example):

$$|2x - 3| \leq 9$$

$$x \in Z \Rightarrow \sum x = ?$$

A)12 B)13 C)14 D)15
E)16

(Solution):

$$-9 \le 2x - 3 \le 9$$

$$-6 \le 2x \le 12$$

$$-3 \le x \le 6 \Rightarrow$$

$$x = -3 - 2 - 1 + 1 + 2 + 3 + 4 + 5 + 6 = 15$$

(Example):

$$\frac{|2x-1|-9}{|x-3|} < 0 \Rightarrow \sum x =?$$

A)0 B)1 C)2 D)3 E)4

(Solution):

$$|x - 3| > 0 \ (and)x - 3 \ne 0, x \ne 0$$

$$|2x - 1| - 9 < 0$$

$$-9 < 2x - 1 < 9$$

$$-8 < 2x < 10$$

$$-4 < x < 5$$

$$-3 \quad -2 \quad -1 \quad 0 \quad 1 \quad 2 \quad 4$$

$$x = -3 - 2 - 1 + 0 + 1 + 2 + 3 + 4 = 1$$

$$a, b \in R^+$$

$$a < |x| < b$$

$$a < x < b \qquad -b < x < -a$$

(Example):

$$3 < |x - 4| < 7 \Rightarrow (SS) = ?$$

A)$(7,11) \cup (-3,1)$ B)$(7,11) \cup (-3,-1)$
C)$(7,12) \cup (-3,1)$

D)$(6,11) \cup (-3,-1)$ D)$(6,11) \cup (-4,1)$

(Solution):

$$3 < |x - 4| < 7$$

$$3 < x - 4 < 7 \qquad -7 < x - 4 < -3$$

$$7 < x < 11 \qquad -3 < x < 1$$

$$(SS) = (7,11) \cup (-3,1)$$

(Example):

$$|x - 2| < |x + 3| \Rightarrow (SS) =?$$

A)$(-\infty, -1)$ B)$(-1,1)$ C)$(-\frac{1}{2}, 1)$

D)$(-\frac{1}{2}, \infty)$ E)$(0,+\infty)$

(Solution):

$$|x - 2| < |x + 5|$$

$$(x - 2)^2 < (x + 3)^2$$

$$x^2 - 4x + 4 < x^2 + 6x + 9$$

$$-5 < 10x$$

$$-\frac{1}{2} < x$$

$$(SS) = (-\frac{1}{2}, +\infty)$$

TEST WITH SOLUTIONS

1. $2x - 4 < 6 \Rightarrow x-?$

A)x>10   B)x>6   C)x>5   D)x<5
E)x<6

(Solution):

$2x - 4 < 6$

$2x < 6 + 4$

$2x < 10$

$\frac{2x}{2} < \frac{10}{2}$

$x < 5$

2. $5 - 2x \le x + 2 \Rightarrow x \ge?$
3.

A)$x \ge -3$ B)$x \ge -2$ C)$x \ge -1$ D)$x \ge 0$
E)$x \ge 1$

(Solution):

$5 - 2x \le x + 2$

$5 - 2 \le x + 2x$

$3 \le 3.x$

$\frac{3}{3} \le \frac{3.x}{3}$

$1 \le x \Rightarrow x \ge 1$

4. $3x - |(2 - x) + 2x| + 1 < x, x - 5 \Rightarrow x <?$

A)x<-6   B)x<-5   C)x<-4   D)x<-3
E)x<-2

(Solution):

$3x - |(2 - x) + 2x| + 1 < x - 5$

$3x - |2 - x + 2x| + 1 < x - 5$

$3x - 2 + x - 2x + 1 < x - 5$

$2x - 1 < x - 5$

$2x - x < -5 + 1$

$x < -4$

5. $\frac{2x-3}{5} < \frac{x-2}{7} + \frac{1}{3} \Rightarrow x <?$
6.

A)$x < \frac{27}{68}$ B)$x < \frac{68}{27}$ C)$x < \frac{27}{50}$ D)$x <$

$\frac{50}{27}$ E)$x < \frac{27}{40}$

(Solution):

$$\frac{2x-3}{5} < \frac{x-2}{7} + \frac{1}{3}$$
$$(21) \quad (15) \quad (35)$$

$$\frac{42x-6}{105} < \frac{15x-30+35}{105}$$

$$42x - 63 < 15x + 5$$

$$42x - 15x < 15x + 5$$

$$27x < 68$$

$$x < \frac{68}{27}$$

7. $\frac{x-1}{4} - 3 \le \frac{2x-1}{3} + x \Rightarrow ? \le x$

A)$-\frac{35}{17} \le x$ 　　　　B)$-2 \le x$ 　　　C)$-\frac{33}{17} \le x$
D)$-\frac{32}{17} \le x$

E)$-\frac{31}{17} \le x$

(Solution):

$$\frac{x-1}{4} - \frac{3}{1} \le \frac{2x-1}{3} + \frac{x}{1}$$
$$(3) \quad (12) \quad (4) \quad (12)$$

146

$$\frac{3x-3-36}{12} \le \frac{8x-4+12x}{12}$$

$$3x - 39 \le 20x - 4$$

$$-39 + 4 \le 20x - 3x$$

$$-35 \le 17.x$$

$$-\frac{35}{17} \le \frac{17.x}{17}$$

$$-\frac{35}{17} \le x$$

8. $4.(x + 1) - 2 > -\frac{1}{2}(x - 1) \Rightarrow x \in ?$

A)x<0   B)x>$\frac{1}{2}$ C)$x > -\frac{1}{3}$ D)$x < \frac{1}{3}$

E)$x < -1$

(Solution):

$$4.(x + 1) - 2 > \frac{x-1}{2}$$

$$4x + 4 - 2 > \frac{1-x}{2}$$

$$\frac{4x+2}{\underset{(2)}{1}} > \frac{1-x}{2}$$

$$\frac{8x+4}{2} > \frac{1-x}{2}$$

$$8x + 4 > 1 - x$$

$$8x + x > 1 - 4$$

$$\frac{9x}{9} > \frac{3}{9}$$

$$x > -\frac{1}{3}$$

9. $x \in R, \frac{-3x+6}{x^2+1} < 0 \Rightarrow x \in?$

A)-1<x<0   B)X>2   C)x<2   D)0<x<$\frac{1}{2}$

E)$-\frac{1}{2} < x < 2$

(Solution):

$$\frac{-3x+6}{x^2+1} < 0$$

$$x^2 + 1 > 0$$

$$-3x + 6 < 0$$

$$6 < 3x$$

$$\frac{6}{3} < \frac{3x}{3}$$

$$2 < x$$

10. $a < b < 0 \Rightarrow |a + b| + |b - a| = ?$
11.

A)2b   B)-b   C)-a   D)-2a

E)2a

(Solution):

$$a < 0, b < 0 \Rightarrow a + b < 0$$

$$a < b \qquad\qquad \Rightarrow 0 < b - a$$

$$|a + b| + |b - a|$$

$$= -a - b + b - a$$

$$= -2a$$

12. $5x + 2 < 3.(x - 1) \Rightarrow x <?$

A)$x < -\dfrac{5}{2}$ B)$x < -3$ C)$x < -\dfrac{7}{2}$

D)x<-4

E)$x < -\dfrac{9}{2}$

(Solution):

$$5x + 2 < 3.(x - 1)$$

$$5x + 2 < 3x - 3$$

$$5x - 3x < -3 - 2$$

$$2x < -5$$

$$\frac{2x}{2} < \frac{-5}{2}$$

$$x < -\frac{5}{2}$$

13. $\frac{-2}{1-x} > 0 \Rightarrow x \in ?$

A)x<0   B)0<x<1   C)x<-1   D)-1<x<0
E)1<x

(Solution):

$$\frac{-2}{1-x} > 0$$

$$1 - x < 0$$

$$1 < x$$

14. $x + 2y - 12 = 0, 2 < y < 6 \Rightarrow ? < x < ?$

A)0<x<6   B)0<x<8   C)2<x<6   D)2<x<6
E)8<x<10

(Solution):

$$x + 2y - 12 = 0$$

$$2y = 12 - x$$

$$\frac{2y}{2} = \frac{12-}{2}$$

$$y = \frac{12-x}{2}$$

$$2 < y < 6 \Rightarrow 2 < \frac{12-x}{2} < 6$$

$$4 < 12 - x < 12$$

$$-8 < -x < 0$$

$$8 > X > 0$$

15. $a < 0 < b \Rightarrow \sqrt{a^2 - 2ab + b^2} + |b - a| = ?$

A)2b   B)2a-b   C)2b-2a   D)2a-2b
E)a-b

(Solution):

$$a < b \Rightarrow a - b < 0$$

$$\Rightarrow b - a > 0$$

$$\sqrt{a^2 - 2ab + b^2} + |b - a| = \sqrt{(a-b)^2} + |b - a|$$

$$= |a - b| + |b - a|$$

$$= -a + b + b - a$$

$$= 2b - 2a$$

16. $|4 - 3x| = 2 \Rightarrow (SS) =?$

A)$\left\{\frac{2}{3}, 2\right\}$ B)$\left\{\frac{1}{3}, 2\right\}$ C)$\left\{\frac{2}{3}, 1\right\}$

D)$\{1,2\}$ E)$\left\{\frac{1}{3}, 1\right\}$

(Solution):

$$|4 - 3x| = 2 \Rightarrow 4 - 3x = 2 \ (or) - 4 + 3x = 2$$

$4 - 2 = 3x$	$3x = 4 + 2$
$2 = 3. x$	$3x = 6$
$\frac{2}{3} = x$	$x = 2$

$$SS = \left\{\frac{2}{3}, 2\right\}$$

17. $1 < x < 3 \quad \Rightarrow \sqrt{x^2 - 6x + 9} - \sqrt{x^2 - 2x + 1} =?$

A)2x-3   B)2x-4   C)2x-5   D)-2x+4

E)-2x+5

(Solution):

$1<x \Rightarrow 0 < x - 1$

$x < 3 \Rightarrow x - 3 < 0$

$\sqrt{x^2 - 6x + 9} - \sqrt{x^2 - 2x + 1} = \sqrt{(x-3)^2} - \sqrt{(x-1)^2}$

$= |x - 3| - |x - 1|$

$= -x + 3 - (x - 1)$

$= -2x + 4$

18. $a < b \Rightarrow a - b < 0$
$$\sqrt{a^2} + \sqrt{(a-b)^2} - \sqrt{(c-b)^2} + \sqrt{(c-a)^2} = ?$$

A)2b-3a   B)2b-a   C)3b-a   D)3b-3a
E)2b-2a

(Solution):

$a < b \Rightarrow a - b < 0$

$b < c \Rightarrow 0 < c - b$

$a < c \Rightarrow 0 < c - a$

$\sqrt{a^2} + \sqrt{(a-b)^2} - \sqrt{(c-b)^2} + \sqrt{(c-a)^2}$

$= |a| + |a - b| - |c - b| + |c - a|$

$= -a + (-a + b) - (c - b) + (c - a)$

$= -a - a + b - c + b + c - a$

153

$= 2b - 3a$

19. $x \in R, |3x - 12| = 12 - 3x \Rightarrow x \leq?$

A)$x \leq 3$ B)$x \leq 4$ C)$x \leq 5$

D)$x \leq 6$

E)$x \leq 7$

(Solution):

$$|3x - 12| = 12 - 3x \Rightarrow 3x - 12 \leq 0$$

$$\Rightarrow 3x \leq 12$$

$$\Rightarrow \frac{3x}{3} \leq \frac{12}{3}$$

$$x \leq 4$$

20. $|5 - x| + x = 7 \Rightarrow (SS) =?$

A)$\{4\}$ B)$\{5\}$ C)$\{6\}$ D)$\{7\}$

E)$\{8\}$

(Solution):

$|5 - x| + x = 7$

$5 - x + x = 7 \quad (Or) - 5 + x + x = 7$

154

$2x = 7 + 5$

$x = 6$

21. $|2x - 1| < 5 \Rightarrow ? < x < ?$

A)-3<x<5   B)-5<x<0   C)-5<x<1   D)-3<x<2

E)-2<x<3

(Solution):

$|2x - 1| < 5 \Rightarrow -5 < 2x - 1 < 5$

$-4 < 2x < 6$

$-2 < x < 3$

22. $|3 - 5x| \leq 13 \Rightarrow ? < x < ?$

A)$-5 \leq x < -2$ B)$-3 \leq x \leq \frac{11}{3}$ C)$-3 \leq x \leq 3$

D)$-2 \leq x \leq \frac{16}{5}$ E)$-2 \leq x \leq 4$

(Solution):

$$|3 - 5x| \leq 13 \quad \Rightarrow -13 \leq 3 - 5x \leq 13$$

$$-16 \leq -5x \leq 10$$

$$\frac{16}{5} \geq x \geq -2$$

23. $|2 - x| > 10 \Rightarrow x \in$?

A)x<-10,x>10   B)x<-6,x>10   C)x<-6,x>12

D)x<-8,x>10   E)x<-8,x>12

(Solution):

$$|2 - x| \geq 10$$

$$2 - x > 10 \qquad (or) \qquad -2 + x > 10$$

$$2 - 10 > x \qquad (or) \qquad x > 10 + 2$$

$$-8 > x \qquad (or) \qquad x > 12$$

24. $x < -1 \quad (and) \qquad 2x + |x + 1| = -6 \Rightarrow x =$?

A)-8   B)-7   C)-6   D)-5   E)-4

(Solution):

$x < -1 \Rightarrow x + 1 < 0$

$2x + 1x + 11 = -6$

$2x - x - 1 = -6$

$x = -6 + 1$

$x = -5$

25. $|2x + 7| < 9 \Rightarrow ? < x <?$

A)-8<x<1   B)-4<x<-5   C)-8<x<8

D)-5<x<6   E)-7<x<9

(Solution):

$$|2x + 7| < 9 \Rightarrow -9 < 2x + 7 < 9$$
$$-16 < 2x < 2$$
$$\frac{-16}{2} < \frac{x}{2} < \frac{2}{2}$$
$$-8 < x < 1$$

QUESTIONS

1. $x < -5, |x + 5| = 3 \Rightarrow x = ?$

A)-2   B)-4   C)-6   D)-7   E)-8

(Solution):

$x < -5 \Rightarrow x + 5 < 0$

$|x + 5| = 3$

$-x - 5 = 3$

$-x = 3 + 5$

$-x = 8$

$x = -8$

2. $|x - 3| + 4 = 5 \Rightarrow x \in ?$

A){-2,4}   B){-6,2}   C){-4,4}   D){6,12}
E){2,4}

(Solution):

$|x - 3| + 4 = 5 \Rightarrow |x - 3| = 5 - 4$

$|x - 3| = 1$

$x - 3 = 1 \qquad\qquad (or) \qquad -x + 3 = 1$

$x = 4 \qquad\qquad (or) \qquad -x = -2$

$\qquad\qquad\qquad\qquad\qquad\qquad\qquad x = 2$

SS={2,4}

3. $-5 < x < -1, |x + 1| + |x + 5| + a = 7 \Rightarrow a =?$

A)1   B)3   C)5   D)7   E)11

(Solution):

$-5 < x \Rightarrow 0 < x + 5$

$x < -1 \Rightarrow x + 1 < 0$

$|x + 1| + |x + 5| + a = 7$

$-x - 1 + x + 5 + a = 7$

$4 + a = 7$

$a = 3$

4. $|x + 1| \le 5 \Rightarrow ? \le x \le?$

A)$-6 \leq x \leq 5$ B)$-6 \leq x \leq 4$ C)$-7 \leq x \leq$

5

D)$-7 \leq x \leq 4$ E)$5 \leq x \leq 6$

(Solution):

$|x + 1| \leq 5 \Rightarrow -5 \leq x + 1 \leq 5$

$-6 \leq x \leq 4$

5. $|3x + 4| < 5 \Rightarrow ? < x < ?$

A)$-6 < x < \dfrac{1}{3}$ B)$-6 < x < 3$ C)$\dfrac{1}{3} < x < 3$

D)$-3 < x < \dfrac{1}{3}$ E)$3 < x < 6$

(Solution):

$|3x + 4| < 5 \Rightarrow -5 < 3x + 4 < 5$

$-9 < 3x < 1$

$\dfrac{-9}{3} < \dfrac{3x}{3} < \dfrac{1}{3}$

$-3 < x < \dfrac{1}{3}$

6. $|2x - a| < a,$ $\qquad a > 0 \Longrightarrow\ ? < x < ?$

A)$0 < x < a$ $\qquad\qquad$ B)$2 < x < a$ $\qquad\qquad$ C)$a < x <$
$2a$

D)$\frac{a}{4} < x < \frac{a}{2}$ $\qquad\qquad$ E)$\frac{a}{2} < x < a$

(Solution):

$|2x - a| < a \Longrightarrow -a < 2x - a < a$

$0 < 2x < 2a$

$\frac{0}{2} < \frac{2x}{2} < \frac{2a}{2}$

$0 < x < a$

7.$a < 0 < b \quad \Longrightarrow |2a - b| + |2b - a| = ?$

A)$3.(b - a)$ $\qquad\qquad$ B)$2(a - b)$ $\qquad\qquad$ C)$b - a$

D)$a - b$ $\qquad\qquad$ E)$a + b$

(Solution):

$a < 0$

$2a < 0$

161

$2a - b < 0$

$0 < b$

$0 < 2b$

$0 < 2b - a$

$|2a - b| + |2b - a| = -2a + b + 2b - a$

$= 3a + 3b$

$3. (b - a)$

$8. x, y \in Z^+, x > y, |y - x| + |y - 1| = 5 \Longrightarrow x = ?$

A)2   B)3   C)4   D)5
E)6

(Solution):

$y < x \Longrightarrow y - x < 0$

$y - 1 \geq 0$

$|y - x| + |y - 1| = 5$

$-y + x + y - 1 = 5$

$x = 6$

9. $b - c > 0, 2b = a + 1, \frac{a}{2} + x = c \Longrightarrow x \in$?

A)$-2, \infty)$

B)$\left(-\infty, -\frac{1}{2}\right)$

C)$\left(\frac{1}{2}, \infty\right)$

D)$\left(-\infty, \frac{1}{2}\right)$

E)$(-\infty, 2)$

(Solution):

$$\frac{a}{2} + \frac{x}{1} = c$$

$$\frac{a+2x}{2} = \frac{c}{1}$$

$$a + 2x = 2c$$

$$b - c > \Longrightarrow 2b - 2c > 0$$

$$a + 1 - (a + 2x) > 0$$

$$a + 1 - a - 2x > 0$$

$$1 > 2x$$

$$x < \frac{1}{2} = \left(-\infty, \frac{1}{2}\right)$$

10. $|x + 7| = 12 \Longrightarrow x_1 + x_2 =$?

A)-16   B)-14   C)-12   D)-10   E)8

(Solution):

$|x + 7| = 12 \Longrightarrow$

$x + 7 = 12 \qquad (or) - x - 7 = 12$

$x_1 = 5 \qquad (or) \qquad - x = 19$

$\qquad\qquad\qquad\qquad x_2 = -19$

$\qquad\qquad\qquad x_{1+}x_2 = 5 + (-19)$

$\qquad\qquad =$-14

11. $x < 0 \Rightarrow |x - 1| + |2 - x| - |x| = ?$

A)1-x   B)3-x   C)1-2x   D)3-2x   E)3-3x

(Solution):

$x < 0 \Longrightarrow x - 1 < 0, 2 - x > 0$

$|x - 1| + |2 - x| - |x| = -x + 1 + 2 - x(-x)$

$\qquad\qquad\qquad\qquad = -x + 1 + 2 - x + x$

$\qquad\qquad\qquad\qquad = 3 - $x

12. $x^2 + y^2 - 2xy - 4 = 0 \Longrightarrow |x - y| = ?$

164

A)-3   B)-1   C)1   D)2   E)4

(Solution):

$x^2 + y^2 - 2xy - 4 = 0$

$(x - y)^2 - 2^2 = 0$

$(x - y + 2).(x - y - 2) = 0$

$\Rightarrow x - y + 2 = 0$ (or) $x - y - 2 = 0$

$x - y = -2$ $x - y = 2$

$= |x - y| = 2$

13. $|x - y + 1| + \sqrt{9x^2 - 6xy + y^2} = 0 \Rightarrow x + y = ?$

A)$\frac{1}{2}$ B)$\frac{3}{2}$ C)1   D)2   E)3

(Solution):

$|x - y + 1| + \sqrt{(3x - y)^2} = 0$

$\Rightarrow |x - y + 1| + |3x - y| = 0$

$\Rightarrow 3x - y = 0, x - y + 1 = 0$

$\Rightarrow y = 3x, x - 3x + 1 = 0$

$-2x + 1 = 0$

$$x = \frac{1}{2}$$

$$y = 3\left(\frac{1}{2}\right)$$

$$= \frac{3}{2}$$

$$x + y = \frac{1}{2} + \frac{3}{2} = 2$$

1. $\dfrac{|1-5|-|-4|+|3|+5}{5+|-4|}=?$

A)0   B)1   C)2   D)3   E)4

2. $|4-\sqrt{8}|+|2-\sqrt{8}|=?$

A)0   B)1   C)2   D)$\sqrt{2}$ E)$2\sqrt{2}$

3. $x<0<y \Rightarrow |x|-|y|-|x-y|-|y-x|=?$

A)y-3x   B)0   C)x+3y   D)x-3y   E)x+y

4. $a<\dfrac{3}{4} \Rightarrow |4a-3|-|12-4a|=?$

A)-10   B)-9   C)-3   D)8   E)12

5. $a<-2 \Rightarrow \sqrt{a^2-7a+18}+\sqrt{a^2+4a+4}=?$

A)4-a   B)a-4   C)2-a   D)2+a   E)a+4

6.$4(x + 1) - 2 < x - 1 \Longrightarrow x \in$?

A)x<0   B)x>$\frac{1}{2}$ C)$x > -\frac{1}{3}$ D)$x < \frac{1}{3}$ E)x<-1

7.$a - \frac{a+4}{3} < 4 \Longrightarrow a \in$?

A)a<0   B)a>8   C)a<$\frac{10}{3}$ D)a>10   E)a<8

8.$|y.(y - 1)| = 2 \Longrightarrow (SS) =$?

A){-2,-1}   B){-1}   C){-1,2}   D){0,2}   E){0,1}

9. $||x - 3| - 4| = 7 \Longrightarrow (SS) =$?

A){-8}   B){-1,2}   C){11}

D){-8,14}   E){14}

10. $|6 - |4x + 3|| = 7 \Longrightarrow (SS) =$?

A) $\left\{-4, \frac{5}{2}\right\}$ B){-4,5}   C) $\left\{-2, \frac{5}{2}\right\}$ D){-4,5}

E){-5,2}

11. $-6 < a < -2$ (and)

$|a + 2| + |a + 6| + x = 10 \Longrightarrow x =?$

A)6   B)a   C)-2a   D)8

E)10

12. $|3 - 2x| = x - 3 \Longrightarrow (SS) =?$

A){0}   B){2}   C)Ø   D){-1}

E){-2}

13. $2a - 4 = |a - 1| \Longrightarrow a =?$

A)0   B)1   C)2   D)3

E)4

14. $|4 - 2x| \le 8 \Longrightarrow (SS) =?$

A)(3,5)   B)(-∞, 6)   C)(-2,+∞)   D)(-2,6)

E)(-2,6)

15. $\left|\frac{x+5}{4}\right| \le 4 \Longrightarrow (SS) =?$

A)$(-\infty, 0)$ B)$(-21,11)$ C)$(-10,21)$

D)$(-2,30)$ E)$(-10,21)$

16. $3 < |2x - 1| < 7$, $x \in Z \Rightarrow \sum x =?$

A)-2 B)-1 C)0 D)1 E)2

17. $\left.\begin{array}{l} x + y = 0 \\ x^2 . y < 0 \end{array}\right\} \Rightarrow |y - x| - |x - 3y| =?$

A)-2y B)4y-2x C)2x-4y D)2x E)2y

18. $|3x - 4| < 5$

How many integer X values are there in the solution set?

A)1 B)2 C)3 D)4 E)5

19. $x < y 0 < z \ (and)$

$|y - x| + |y - z| + 2.|x - z| = 3 \Rightarrow z - x =?$

A)1   B)2   C)3   D)4   E)5

$20.1 \leq |2x - 1| < 5 \implies (SS) =?$

A)(-1,0)U(2,3)   B)(1,3)U(-2,1)   C)(1,3)U(0,2)   D)(1,3)U(-2,1)   E)(-1,1)U(1,3)

21. $|3 + |x^2 + 4|| = 23 \implies \sum x =?$

A)0   B)2   C)5   D)9   E)14

$22.2. \frac{x^4}{|x^3|} + 3x = 20 \implies \sum x =?$

A)4   B)8   C)12   D)16   E)24

23. $|3 + 2x| = 15 \implies \sum x =?$

A)9   B)6   C)3   D)0   E)-3

(ANSWERS)

1.B	2.C	3.D	4.B	5.A	6.E
7.E	8.C	9.D	10.A	11.A	12.C
13.D	14.E	15.B	16.D	17.E	18.C
19.A	20.D	21.A	22.E	23.E	

1. $-4 < x < 2 \Longrightarrow |x - 2| + |x + 4| + 1 = ?$

A)3　　　　B)4　　　　C)5　　　　D)6　　　　E)7

2. $x < 0 < y < -z \Longrightarrow |x - y| - |y - z| - |x + z| = ?$

A)x-2　　　　B)0　　　　C)2y　　　　D)2z　　　　E)2(x+y+z)

3. $x < 0 \Longrightarrow \big||x| + 2\big| - |-x| = ?$

A)2　　　　B)x　　　　C)2x+2　　　　D)2-x　　　　E)2x-2

4. $x < 0 < y \Longrightarrow \sqrt{x^2} - |x - y| + \sqrt{y^2} = ?$

A)-2x　　　　B)-2y　　　　C)x-y　　　　D)2(x+y)　　　　E)0

5. $|2x - 3| = 5 \Longrightarrow (SS) = ?$

A){4}　　　B){-1}　　　C){-1,4}　　　D){1,4}　　　E){-1,2}

6. $\big||2x + 1| + 4\big| = 3 \Longrightarrow (SS) = ?$

A)(-1)   B)(-4,-1)   C)(-1,1,4)   D)∅   E)(-4,1.4)

7. $x < y < 0 < z \Longrightarrow$

$$\sqrt{x^2 - 2xy + y^2} - \sqrt{x^2 - 2xz + z^2} + \sqrt{y^2 - 2yz + z^2} =?$$

A)0   B)3x   C)y   D)x+y+z   E)z

8. $A = \left\{ x\sqrt{(x - 3^2} = 3 - x, x \epsilon R \right\} \Longrightarrow x \in$?

A)(0,3)   B)(-∞, 3)   C))3,+∞)   D)(-3,3)   E)R

9. $|-3x + 7| = 5 \Rightarrow \sum x =$?

A)6   B)$\frac{4}{3}$ C)$\frac{8}{3}$ D(12   E)$\frac{14}{3}$

10. $x \in R, x < -2 \Longrightarrow$

$$x + \sqrt{x^2 + 5x + 1 + \sqrt{x^2 - 6x + 9}} =?$$

A)x-2   B)x   C)=x-2   D)2-x   E)-2

11. $2 < x < 6 \Rightarrow |x - 2| + |x - 6| + 2x = 10 \Rightarrow x =?$

A)3   B)4   C)5   D)6   E)8

12. $|x| - 2 = |x - 2| \Rightarrow (SS) =?$

A)R   B)$(-\infty, 0)$ C)$(-2, +\infty)$ D)$(2, +\infty)$
E)$\emptyset$

13. $\left|3 - |x - 1|\right| = 2 \Rightarrow \sum x =?$

A)0   B)2   C)3   D)4
E)5

14. $|2x - 6| + |4y + 20| = 0 \Rightarrow x + y =?$

A)-2   B)-1   C)3   D)8
E)14

15. $|x| + |y| + |z| = 2, \quad x, y, z \in Z$

A)-1   B)0   C)2   D)4   E)6

16. $x > 2, \left|3x + |2 - x|\right| = 10 \Rightarrow (SS) = ?$

A){2}   B){3}   C){4}   D){-2,2}   E){2,3}

17. $a < b < 0 < c \Rightarrow \dfrac{|a-b|+|c-b|}{|-c|+|-a|} = ?$

A)1   B)2   C)-1   D)a   E)2a-b

18. $|2x - 4| + |y - 5| + |z + 2| = 0$

$\Rightarrow x + y + z = ?$

A)2   B)3   C)4   D)5   E)6

19. $x < -5 \Rightarrow \left|5x + |4x - 5|\right| + x = ?$

A)-5   B)-8x+5   C)-2x-5   D)-x   E)-2x+5

20. $3 < x < 4 \Rightarrow \sqrt{x^2 - 5x + 5} + \sqrt{x^2 - 8x + 16 +}\ x = ?$

176

A)2x-3   B)-2x+3   C)5   D)3   E)4x+2

21. $\sqrt{(2-|x|)^2} = 1 \Longrightarrow \sum x^2 =?$

A)74   B)52   C)50   D)34   E)20

22. $|x|^{(x^2+x-2)} = 1 \Longrightarrow (SS) =?$

A){-2,1}   B){2,-1}   C){-2,0,2}   D){2}   E){-1}

23. $|3x-2| = 2x-1 \Longrightarrow (SS) =?$

24. $\frac{x+3}{|x-1|+2} = \frac{1}{2} \Longrightarrow (SS) =?$

A)(-5,-1)   B)(-5)   C)(-1)   D)Ø   E)(-1,+5)

(ANSWERS)

1.E	2.D	3.A	4.E	5.C	6.D
7.A	8.B	9.E	10.E	11.A	12.D
13.D	14.A	15.B	16.B	17.A	18.D

19.A   20.A   21.E   22.C   23.C   24.C

1.$x < 0 < y \Rightarrow |x| - |y| - |x - y| - |y - x| =?$

A)-3x+y   B)0   C)x+3y   D)x-3y   E)x+y

2.$|3 - 2\sqrt{3}| + |4 - 2\sqrt{3}| =?$

A)0   B)1   C)2   D)$\sqrt{2}$ E)$2\sqrt{2}$

3.$3.|-3 + 4 - (-2)| - 4.|-3| =?$

A)-3   B)-2   C)-1   D)0   E)1

4.$x < 3 \Rightarrow |x - 3| + 5x - 4 =?$

A)6x-1   B)6x-7   C)4x+2   D)2x+1   E)4x-1

5.$a < \frac{1}{2} \Rightarrow |1 - 2a| - |-2a + 6| =?$

A)-10   B)-9   C)-5   D)8   E)12

6. $x > 0, y < 0 \Rightarrow \sqrt{x^2}+\sqrt{y^2}=?$

A)2x   B)x-y   C)x+y   D)2y   E)0

7. $a < -2 \Rightarrow \sqrt{a^2+4a+4}=?$

A)4-a   B)-2-a   C)2+a   D)a+4   E)a-4

8. $1 < a < 2 \Rightarrow \sqrt{a^2-4a+4}-\sqrt{a^2-2a+1}+1=?$

A)5-a   B)3-a   C)2(a+2)   D)4-a   E)2(2-a)

9. $a < 0, b > 0 \Rightarrow \sqrt{9a^2}-\sqrt{4b^2}-|a-b|=?$

A)-(a+b)   B)-(2a+3b)   C)3a-b   D)2(a-b)   E)-2(2a-b)

10. $a < 0 \Rightarrow \frac{|-5a|+|a|+|-4a|}{|a|}=?$

A)10a   B)-12a   C)15   D)-10   E)10

11. $|2-4x|=6 \Rightarrow (SS)=?$

A){-8}   B){-2,1}   C){1}   D){-1,2}   E){14}

12. $|5x - 3| = x - 9 \Longrightarrow (SS) =?$

A)$\left\{\frac{3}{2}\right\}$ B){2}   C){3}   D)$\left\{-\frac{3}{2}\right\}$ E)$\emptyset$

13. $|x - 1| - |x + 1| = 1 \Longrightarrow (SS) =?$

A){0}   B)$\left\{-\frac{1}{2}\right\}$ C)$\left\{-\frac{2}{3}\right\}$ D){2}   E){3}

14. $\left||4x + 3| + |5 - x|\right| = 7 \Longrightarrow x =?$

A)-2   B)-1   C)0   D)2   E)9

15. $n \epsilon Z^+, x < 0 < y \Longrightarrow \sqrt[2n]{x^{2n}} + \sqrt[2n-1]{y^{2n-1}} =?$

A)x-y   B)y-x   C)-x-y   D)x+y   E)xy

16. $\dfrac{1}{|x+1|+|4+4x|} = \dfrac{4}{5} \Longrightarrow x_1 . x_2 =?$

A)5 B)$\frac{12}{25}$ C)$\frac{15}{16}$ D)8 E)12

17. $|6 - |4x + 3|| = 7 \Rightarrow \sum x =?$

A)-4 B)3 C)$-\frac{2}{7}$ D)$\frac{4}{5}$ E)$-\frac{3}{2}$

18. $\frac{x+5}{|x-5|-3} = 2 \Rightarrow (SS) =?$

A)$\left\{-\frac{1}{3}, 5\right\}$ B)$\{1,21\}$ C)$\left\{-\frac{1}{3}, 21\right\}$ D)$\left\{\frac{1}{3}, \frac{2}{5}\right\}$
E)$\{3\}$

19. $|x^2 + 3| = |x + 5| \Rightarrow \sum x =?$

A)-2 B)0 C)1 D)2 E)5

20. $|3x - 5| = -x \Rightarrow (SS) =?$

A)$\{1\}$ B)$\left\{\frac{5}{2}\right\}$ C)$\left\{\frac{4}{3}\right\}$ D)$\emptyset$ E)$\left\{\frac{2}{3}\right\}$

21. $|x - b| = |4x + b| \Rightarrow \sum x =?$

A)-3b   B)2b   C)$-\dfrac{2b}{3}$ D)$\dfrac{b}{2}$ E)$-\dfrac{5}{2}$

22. $|a^2 - 16| - |a + 4| = 0 \Rightarrow (SS) =?$

A){-2,4,3}   B){-4,3,5}   C){-2,3,6}   D){2,-1,0}   E){-1,0,1}

23. $a < 0 < b, b > |a| \Rightarrow$

$$\frac{|(a-b)(a+b)|-\sqrt{a^2-2ab+b^2}}{a-b} =?$$

A)a-b+1   B)1-a-b   C)1+a   D)-1+a+b

E)a+b+1

(ANSWERS)

Z	2.B	3.A	4.E	5.C	6.B
7.B	8.E	9.B	10.E	11.D	12.E
13.B	14.B	15.B	16.C	17.E	18.C

19.C   20.D   21.C   22.B   23.B

1. $|x - 5| < |x - 3|, x \in Z \Longrightarrow x_{min} =?$

A)4   B)5   C)6   D)7   E)10

3. $|4x - 2x| \leq 8 \Longrightarrow x \in?$
4.

A)(3,5)   B)$(-\infty, 6)$ C)$(-2, +\infty)$ D)(-2,6)
E)(-2,6)

5. $\left|\frac{4x-3}{5}\right| > 3 \Longrightarrow x \in?$
6.

A)$\left(\frac{9}{5}, 5\right)$ B)$\left(-3, \frac{9}{2}\right)$ C)$-\infty, \frac{9}{2}$
D)$(-\infty, -3)U\left(\frac{9}{2}, +\infty\right)$ E)$(-\infty, -3)U\left(\frac{9}{2}, +\infty\right)$

7. $|2x + 1| \leq |2x - 3|, x \in Z \Longrightarrow x_{max} =?$
8.

A)0   B)1   C)2   D)3
E)4

5. $x \in Z, 5 < |x - 4| < 8 \Longrightarrow \sum x =?$

A)-5   B)16   C)21   D)25
E)30

6. $\left|\frac{x+7}{2}\right| \leq 7 \implies x \in$?

A)$(-\infty, 0)$ B)(-21,7)   C)(-10,21)   D)(-2,30)
E)(-10,21)

7. $|x-2| \leq |x-4| \implies x \in$?
8.

A)$(-\infty, 0)$ B)$(-\infty, -3)$ C)$(-\infty, -3)$
D)$(-\infty, 3)$ E)(2,3)

8. $x \epsilon Z$,

$3 < |2x - 1| < 7 \implies \sum x =$?

A)-2   B)-1   C)0   D)1
E)2

9. $|2x - 2| \leq |x + 2| \implies x \in$?

A)(-2,0)   B)(0,4)   C)(-2,4)   D)(-
2,0)∪(2,+∞)   E)(4,+∞)

10. $\left.\begin{array}{l}|x + 2| < 4 \\ |x - 1| > 3\end{array}\right\} \Rightarrow \Sigma x =?$

A)6   B)7   C)-9   D)-12

E)-14

11. $x. |x| \leq 2 \Rightarrow (SS) =?$

A)$(-\infty, -2)$ B)$(\sqrt{2}, \infty)$ C)$(-\infty, -\sqrt{2})$ D)$(-\sqrt{2}, \sqrt{2})$

E)$(-\infty, \sqrt{2})$

12. $x \in Z$

$4 < |2x - 1| \leq 5 \Sigma x =?$

A)0   B)1   C)2   D)3   E)4

13. $|x + 1| \leq |x - 3| \Rightarrow (SS) =?$

A)$(0,1)$ B)$(-\infty, 1)$ C)$(1,+\infty)$ D)R   E)$(-1,1)$

14. $x \in Z$

$2 \le |x - 1| \le 5 \Rightarrow \sum x = ?$

A)4   B)5   C)6   D)7   E)8

15. $\left| \frac{3}{x-2} \right| \ge 1 \Rightarrow (SS) = ?$

A)(-1,5)-(2)   B)(0,2)   C)(2,3)   D)(2,+∞)   E)(-∞, 3)

16. $x \in Z$

$\sqrt{x^2 - 6x + 9} = 3 - x, \ |x + 2| = x + 2$

$\Rightarrow \sum x = ?$

17. $|3x - 2| < 8 \Rightarrow (SS) = ?$

A) $(-2,0)$ B)$(-2,\frac{10}{3})$ C)$(0,2)$ D)$(-2,\frac{10}{3})$
E)$(-2,\frac{10}{3})$

18. $|x - 2| < 4 \Rightarrow (SS) = ?$

A)-2<x<6   B)-2<x<0   C)2<x<6   D)-2<x<2
E)-2<x<4

19. $1 < \frac{|1-x|}{4} < 2 \Longrightarrow x \in$?

A)(-9)   B)(-4,7)   C)(-7,-3)∪(5,9)   D)(5,12)

20. $|x^2 - 8| < 8 \Longrightarrow x \in$?

A)(-4,4)   B)(-4,4)-(0)   C)R   D)(-∞,4)
E)(-4,+∞)

21. $|x^2 - 2| \leq 7 \Longrightarrow x \in$?

A)(-4,0)   B)(-3,3)   C)(-2,4)   D)(0,4)
E)3,5)

22. $|x^2 + 3| \leq 7 \Longrightarrow x \in$?

A)(-2,2)   B)(-∞, 2)   C)(-2,+∞)   D)R-(-2,2)
E)R-(-√3, √3)

$23. x^2 + 2|x| - 15 < 0 \Rightarrow x \in ?$

A)(-5,3) B)-3,5) C)(-3,3) D)(-3,+∞)
E)(-∞, −5)

(ANSWERS)

1.B	2.E	3.B	4.A	5.B	6.B
7.D	8.D	9.B	10.B	11.E	12.B
13.B	14.E	15.A	16.B	17.D	18.A
19.C	20.B	21.B	22.A	23.C	

$1.4^{|x-2|} = 64 \Longrightarrow (SS) = ?$

A)(-1,5)   B)(-1)   C)(5)   D)(-5,1)   E)(2)

$2. |2x + 4| = 7 \Longrightarrow (SS) = ?$

A)$\left\{\frac{11}{2}, \frac{-3}{2}\right\}$ B)$\left\{\frac{3}{2}, \frac{11}{2}\right\}$ C)$\left\{-\frac{11}{2}, \frac{2}{3}\right\}$ D)$\left\{-\frac{11}{2}, \frac{3}{2}\right\}$

E)$\left\{\frac{5}{2}, 6\right\}$

$3. |2x + 6| = |2x - 2| \Longrightarrow x = ?$

A)0   B)-1   C)3   D)6   E)24

$4. |x - 2| + |x - 3| = 7 \Longrightarrow (SS) = ?$

A)(0,1,2)   B)(1,6)   C)(-1,6)   D)(-1)   E)(6)

$5. x \in Z, |2x - 9| \leq 5 \Longrightarrow \sum x = ?$

A)22   B)23   C)25   D)26   E)27

6. $a < 0 < b \Rightarrow$

$|a - b| - |b - 2a| - |a| =?$

A)b   B)-b   C)2b-a   D)2a   E)-a

7. $|x + y + 3| + |x - y - 1| = 0 \Rightarrow |y| =?$

A)1   B)2   C)3   D)4   E)5

8. $|2x + 1| - |x - 2| = 0 \Rightarrow \sum x =?$

A)0   B)-1   C)1   D)2   E)$-\frac{8}{3}$

9. $a < |a| = b,$

$|a - b| - |a + b| + |a + b| = 0 \Rightarrow a =?$

A)-4   B)-2   C)0   D)2   E)4

10. $a < 0 < b \Rightarrow |a| - |-b| + |a - b| + |b + 2| =?$

A)-b+2   B)-2a+b+2   C)-b-a-2   D)a-b-2   E)-2b+2

11. $|x^2 - 4x| < x \Longrightarrow ? < x <?$

A)0<x<3   B)2<x<3   C)3<x<5   D)4<x<5
E)4<x<6

12. $x \in Z, 4 \leq |x - 2| < 5 \Longrightarrow \sum x =?$

A)6   B)5   C)4   D)-8   E)-2

13. $a^2 < a$

$a.c < 0 \Longrightarrow |a| - |c - a| + |c| =?$

A)-2c   B)0   C)a   D)-2c   E)2a-2c

14. $2 < x < 5 \Longrightarrow \frac{|x-5|-|x-2|}{|x|-|x+1|} =?$

A)x+7   B)5-x   C)2x-7   D)-x+2   E)2x+7

15. $0 < x < 3 \Longrightarrow |x - 3| + \sqrt{x^2 - 6x + 9} - |3 - 0| =?$

A)3-x   B)x-3   C)3-2x   D)4x-3   E)3

16.$x < 0 < y \Longrightarrow$

$$\frac{|x-2y|-|y-x|-|y|}{|y|-|x|} =?$$

A)-x   B)2y   C)-2   D)1   E)0

17.$X > y, x+y=0$

$$\frac{|x^2-2xy|+|y-2x|}{2|x|+|y|} = 4 \Longrightarrow (x - y) =?$$

A)-1   B)0   C)3   D)6   E)7

18.$x \in Z, \frac{|x|+4}{1-|2x-3|} < 0 \Longrightarrow \sum x =?$

A)-4   B)-3   C)0   D)6   E)8

19.$3^x = 220 \Longrightarrow |x - 5| + |x - 4| =?$

A)0   B)1   C)2   D)9   E)4

20.$x \in Z, 2 \leq |x - 1| < 6 \Longrightarrow \sum x =?$

A)0   B)2   C)5   D)8   E)12

21. $x \in Z, \left|\frac{2^x}{8} - 2\right| \le 7 \Longrightarrow \max(x) =?$

A)0   B)1   C)4   D)6   E)7

22. $a < 0 < b \Longrightarrow |a| + |b| - |a - b| =?$

A)-a   B)-b   C)0   D)a   E)b

23. $a < b < 0 \Longrightarrow$

$|a + b| - |b - a| + \sqrt{a^2 - 2ab + b^2} =?$

A)2a   B)a+b   C)b-a   D)a-b   E)-a-b

24. $|2x - 1| = 5 \Longrightarrow (SS) =?$

A)(-3)   B)(-2)   C)(3)   D)(-2,3)   E)(-3,2)

25. $|3 - 2x| + x = 6 \Longrightarrow (SS) =?$

A)(-3)   B)(-2)   C)(3)   D)(-3,3)   E)(1,3)

26. $|3x - 6| + |2 + y| + |9 + 3z| = 0 \Rightarrow x + y + z =?$

A)-3   B)-2   C)-1   D)3   E)5

27. $\left|\frac{3}{x} - 1\right| + 3 = 0 \ (SS) =?$

A)$\emptyset$ B)(3)   C)1,3)   D)(-1,3)   E)R

28. $\left|\frac{x-5}{7}\right| > 0 \Rightarrow (SS) =?$

A)(5+∞)   B)(-5,+∞)   C)(-5,5)   D)R   E)R-(5)

(ANSWERS)

1.A	2.D	3.B	4.C	5.E	6.D
7.B	8.E	9.B	10.B	11.C	12.C
13.B	14.C	15.C	16.E	17.B	18.B
19.B	20.D	21.D	22.C	23.E	24.D
25.D	26.A	27.A	28.E		

1. $x < -2 \Rightarrow |2x| + |x + 2| - |-2x| - |-x| = ?$

A)-2   B)1   C)x   D)2x   E)5

2. $2(x - 1) \leq -1 + 3(x + 2) \Rightarrow (SS) = ?$

A)$x \geq 3$ B)$x \geq 1$ C)$x \leq -1$ D)$x \geq -3$
E)$x \geq -7$

3. $\frac{3}{4}(x + 8) \geq -\frac{1}{2}(x + 2) \Rightarrow (SS) = ?$

A)$x \geq -2$ B)$X > \frac{8}{5}$ C)$x > -1$ D)$x > \frac{1}{2}$
E)$x \geq -16$

4. $\frac{5}{6}(x - 12) \geq 1 + \frac{1}{6}(x - 6) \Rightarrow (SS) = ?$

A)$(6, +\infty)$ B)$(9, +\infty)$ C)$(15, +\infty)$ D)$(-\infty, 21)$
E)$(-\infty, 18)$

5. $a < 0 < b \Rightarrow |a - b| + |-b| - |-a| = ?$

A)a   B)2a   C)2b   D)2a+2b
D)c-b

6. $a < b < 0 \Rightarrow |a + b| + |b - a| + |a - b| = ?$

A)2a   B)2b   C)b-3a   D)a-2b
E)2a+3b

7. $\frac{x+7}{x} > -\frac{3}{4} \Rightarrow (SS) = ?$

A)$(-\infty, -4)$ B)$(-\infty, -3)$ C)R(-4,-2)   D)$(-\infty, -1)$
E)R(-4,0)

8. $\left|x - \frac{2}{3}\right| = \frac{2}{3} - 1 \Rightarrow (SS) = ?$

A)(1)   B)(2)   C)(3,6)   D)$\emptyset$
E)(4,5)

9. $|x| = 3x - 8 \Rightarrow (SS) = ?$

A)(2)   B)(4)   C)(5)   D)(6)
E)(8)

10. $|x + 1| = 2x + 4 \Rightarrow (SS) =?$

A)(0)   B)$\left\{-\frac{1}{2}\right\}$ C)$\left\{-\frac{5}{3}\right\}$ D)(-2)   E)(-3)

11. $|6 - x| = x \Rightarrow (SS) =?$

A)(1)   B)(2)   C)(3)   D)(4)   E)(5)

12. $|2x + 3| = 9 \Rightarrow (SS) =?$

A)(-2,2)   B)(-4,3)   C)(-6,2)   D)(-6,3)   E)(-7,5)

13. $\left|\frac{3x}{5} + \frac{x}{2}\right| = 1 \Rightarrow (SS) =?$

A)$\left\{-\frac{10}{11}, \frac{10}{11}\right\}$ B)$\left\{-\frac{12}{13}, \frac{12}{13}\right\}$ C)$\left\{-\frac{1}{2}, \frac{1}{2}\right\}$

D)$\left\{\frac{3}{5}\right\}$ E)$\frac{2}{5}$

14. $|3x - 4| = 6 - x \Rightarrow \sum x =?$

A)$\frac{7}{2}$ B)$\frac{5}{2}$ C)$\frac{3}{2}$ D)$\frac{1}{2}$ E)0

15. $|5x + 2| = |2x + 5| \Longrightarrow \sum x =?$

A)9   B)8   C)2   D)1   E)0

16. $|4x - 6| + x^2 = 0 \Longrightarrow (SS) =?$

A)(1)   B)(2)   C)(4)   D)(6)   E)$\emptyset$

17. $x < y < 0 \Longrightarrow \sqrt{16x^2} + \sqrt{25y^2} - |4x + y| =?$

A)-2x   B)-2x+2y   C)-4y   D)-4y+x
E)x+y

18. $a < 0 < b \Longrightarrow \dfrac{\sqrt{a^2}+\sqrt{b^2}}{|a|+|-b|} =?$

A)-1   B)1   C)2   D)3   E)a+b

19. $|x - 1| + |x - 2| = 13 \Longrightarrow \sum x =?$

A)-3   B)-2   C)0   D)2   E)3

20. $x < 2 \implies \sqrt{x^2 - 4x + 4} - |2 - x| =?$

A)6x   B)-4x   C)$-\frac{7}{2}$ D)-3   E)0

21. $x < 0 \implies \left| x - |3x| + |-x| \right| =?$

A)x   B)-x   C)-2x   D)-3x   E)-4x

22. $x < 1 \implies \sqrt{x^2 - 3x + \sqrt{x^2 - 2x + 1}} =?$

A)2-x   B)x-2   C)$|-x|$ D)x   E)2

23. $|3x - 2| < 5 \implies (SS) =?$

A)$\left(-1, \frac{7}{3}\right)$ B)$\left(-1, \frac{7}{3}\right)$ C)$(-\infty, -1)$ D)$\left(\frac{7}{3}, +\infty\right)$

$E) \emptyset$

24. $\left| |x + 3| + 2 \right| = 5 \implies \sum x =?$

A)-10   B)-6   C)-3   D)0
E)3

(ANSWERS)

1.A	2.E	3.E	4.C	5.C	6.C
7.E	8.D	9.B	10.C	11.C	12.D
13.A	14.C	15.E	16.E	17.C	18.B
19.E	20.E	21.D	22.A	23.A	24.B

1. $a < b < c < 0 \Rightarrow$

$|b - c| - |b + a| + |a - c| = ?$

A)2b   B)b   C)0   D)2c   E)a-b

2. $a < 0 \Rightarrow \dfrac{|4a-3|}{|-3a|+|3-a|} = ?$

A)$\dfrac{4a-3}{-2a-3}$ B)$\dfrac{4a-3}{2a+3}$ C)1   D)$\dfrac{4a-3}{4a+3}$

$E) -1$

3. $x < 0 \Rightarrow \left|4x - \left|2 + |-2x|\right|\right| + \left|2x - |x|\right| = ?$

A)5x+2   B)x   C)-x+2   D)3x+2

E)-9x+2

4. $a \in R, 4 < a < 7$

$f(x) = |a + x - 4| - |x - a| \Rightarrow f(4) = ?$

A)4   B)5   C)6   D)7

E)8

5. $x < 0 < y \Rightarrow$

$$\sqrt{x^2 - 6xy + 9y^2} - 3y\sqrt{4x^2} - |2y - x| - 6xy = ?$$

A)y-x B)y C)-2y+x D)-x
E)0

6. $a < -2 \Rightarrow \left|4a - |3a|\right| - 2 = ?$

A)0 B)a-2 C)a+4 D)6
E)-7a-2

7. $x, y \in R$

$|x + 2y| = 2$

$f(x, y) = \sqrt{x^2 + 4xy + 4y^2} - (x + 2y)^3$

$\Rightarrow \max\{f(x, y)\} = ?$

A)12 B)10 C)4 D)-4
E)-6

8. $a > 0 \Rightarrow \sqrt[5]{a^5} - \sqrt{(-a)^2} = ?$

A)2   B)1   C)0   D)-1   E)-
2

$9. a < b < 0 \Rightarrow |a + b| - |b| - |-a| = ?$

A)2b   B)-2b   C)a-b   D)0   E)-2a

$10f(x) = |4x + 5|,$

$f(x) < 0 \Rightarrow (SS) = ?$

$A)(0,1)$ $B)(2,3)$ $C)(0,1)$ $D)R$

$11. 5^{2x} - 26.5^x + 25 \leq 0 \Rightarrow (SS) = ?$

A)(0,2)   B)(0,1)   C){}   D)(0,1)
E){2}

$12. \frac{-6}{1+|-x|} < 0 \Rightarrow (SS) = ?$

A)(-1,0)   B)(0,1)   C)R   D)∅
E)(1,∞)

13. $(x - 3)^{(x-3)} = 1 \Longrightarrow (SS) =?$

A)(2,3)   B)(3,4)   C)(2,3,4)   D)(4)
E)Z^+

14. $-3 < |2x - 1| \leq (SS) =?$

A)(-3,5)   B)(-1,3)   C)(-1,2)   D)(-2,3)   E)(-2,3)

(ANSWERS)

1.A	2.C	3.E	4.A	5.B	6.E
7.B	8.C	9.D	10.E	11.D	12.C
13.D	14.C				

Polynomials

Definition

A polynomial is any function f of the form

$$P(x) = a_0.\,a_1x.\,a_2x^2 \ldots\ldots\ldots\ldots a_nx^n \text{ where}$$

$a_0.\,a_1.\,a_2 \ldots\ldots\ldots\ldots a_n$ are real numbers ($\in R$) and where n

Is a natural number.

Example:

1.$P(x)$=$4x^6 - 3x^{-2} + 6x+1$

$P(x)$ is not a polynomial because the power of

-$3x^{-2}$ is -2,but -2 is not a natural number.

2. $P(x) = 6x^5 + 7x^3 + x^{\frac{1}{2}} + 3$

$P(x)$ is not a polynomial because of the power "$\frac{1}{2}$",

"$\frac{1}{2}$" is not a natural number.

Example:

($P(x)$ is a polynomial)

$P(x) = ax^6 + (a\text{-}b\text{+}3)x^{-3} + (a + b - 9)x^{-2} + bx$

$\Rightarrow P(x) =?$

A)$2x^6 + 3X$ 4X

B) $3x^6 + 6X$

C) $5x^6 +$

D) $2x^6 + 5X$

E) $3x^6 + 9X$

(Solution):

$a-b+3=0$
$\underline{+a+b-9=0}$
$2a-6=0$

a=3,(and)b=6

$\Rightarrow P(x) = 3x^2 + 6$

-Answer B

(Example)

(P(x) is a polynomial)

$P(x^2)=(a\text{-}2)x^5 + ax^4 + (b - 4)x^3 + 2bx^2 + 3b$

$\Rightarrow P(x) =?$

A)$2x^2+8X+12$ $4x^2+12X+6$

B) $2x^2+4X+1$

C)

D) $3x^2+5X+8$

E) $2x^2+6X+4$

(Solution):

208

$$P((\sqrt{x})^{-2}) = (a\text{-}2)(\sqrt{x})^5 + a(\sqrt{x})^{-4} + (b-4)(\sqrt{x})^3 + 2b(\sqrt{x})^2 + 3b$$

$$\Rightarrow P(x) = (a-2)x^{\frac{5}{2}} + ax^2 + (b\text{-}4)x^{\frac{3}{2}} + 2bx + 3b$$

a=2=0$\Rightarrow a = 2$

b-4=o$\Rightarrow b = 4$

P(x)=$2x^2 + 2.4x + 3.4$

$=2x^2 + 8x + 12$

-Answer

A

(Example):

$$x^3 \cdot P(x) = ax^8 + (b-2)x^5 + (a-4)x + b - 5 \Rightarrow P(x) =?$$

A)$3x^5 + 2x^2$ B) $3x^2 + 4x^2$ C) $4x^5 + 3x^2$

D) $8x^5 + 2x^2$ E) $7x^5 + 2x^2$

(Solution):

$$\frac{x^3 P(x)}{x^3} = \frac{ax^6}{x^3} + \frac{(b-2)x^5}{x^3} + \frac{(a-4)x}{x^3} + \frac{b-5}{x^3}$$

P(x)=$ax^5 + (b-2)x^2 + (a\text{-}4)x^{-2} + (b-5)x^{-3}$

a-4=0 (and) b-5=0

a=4,b=5

P(x)=$4x^5 + (5-2)x^2$

$$=4x^5 + 3x^2$$

-Answer C

(Example):

$$P(x)=-3x^3 + 4x^2 - X + m + 1$$

$$P(2)=4 \Rightarrow m =?$$

(Solution):

$$P(2)=-3.2^3 + 4.2^2 - 2 + m + 1 = 4$$

-3.8+4.4-2+m+1=4

-24+16-2+m+1=4

-9+m=4

m=9+4

m=13

(Example):

$$P(x)=5x^6 - 4x^3 + 11 \Rightarrow \qquad P(\sqrt[3]{2}) =?$$

(Solution):

$$P(\sqrt[3]{2}) =5.(\sqrt[3]{2})^6 - 4(\sqrt[3]{2})^2 + 11$$

=5.4-4.2+11

=20-8+11

=23

(Example):

$n \in Z$

$P(x)=5.(x - 2)^{2n} - 7(2 - x)^{2n-1} \Rightarrow P(1) =?$

(Solution):

$P(1)=5.(1 - 2)^{2n} - 7(2 - 1)^{2n-1}$

$= 5.(-1)^{2n} - 71^{2n-1}$

=5.1-7.1

=5-7

=-2

EQUALITY OF POLYNOMIALS

$P(X)=a_n x^n + a_{n-1} x^{0-1} + \cdots + a_1 x + a_0$

$Q(x)= b_n x^n + b_{n-1} x^{n-1} + \cdots + b_1 x + b_0$

$P(x)=Q(X) \Rightarrow a^n = b_n, a_{n-1}, \dots. a_{1=b_1, a_0=b_0}$

(Example):

$P(x)=(x^2+1).(x+3)$

$Q(x)=x^3 + ax^2+bx+c$

$P(x)=Q(x) \Rightarrow a =?, b =?, c =?$

(Solution):

$P(x)=(x^2 + 1)(x + 3) = x^3 + ax^2 + bx + c = Q(x)$

$P(x)=x^3 + 3x^2 + 1x + 3 = |x^3 + ax^2 + bx + c = Q(x)$

$\Rightarrow a = 3, b = 1, c = 3$

(Example):

$P(x)=(x-2)(x^2 + px + 3) + x - 5$

$Q(x)=x^3 + 3x^2 + bx + c$

$P(x)=Q(x) \Rightarrow b + p =?$

A)-5   B)-4   C)-3

D)-2   E)-1

(Solution):

$P(x)=x^3 + x^2(p - 2) + (4 - 2p)x - 11$

$Q(x)=x^3 + 3x^2 + bx + c$

$P(x)=Q(x) \Rightarrow p - 2 = 3 \Rightarrow p = 5, 4 - 2p = b \Rightarrow b = 4 - 10 = -6, c = -11$

b+p=5-6=-1

SUM OF COFFICIENTS ON POLYNOMIALS

A polynomial P(x) is given

1.To find the sum of the coefficients of P(x).write 1

Instead of x.

2.To find the constant term of P(x).write 0 instead of x.

P(0)=Constant term

(Example):

$P(x)=(x^2 - x + 2)^3(x^4 - 2x + 4)^2$

What is the sum of the coefficients of P(x)?

A)48   B)66   C)72   D)84   E)90

(Solution):

X=1

$P(1)=(1 - 1 + 2)^3.(1 - 2 + 4)^2$

$=(2)^3.(3)^2 = 8.9 = 72$

                                             -Answer C

(Example):

$P(x^3 + 8) = x^6 - 2x^3 + 1.$

What is the constant term of P(x)?

A)70　　　　　　　　B)72　　　　　　　C)78　　　D)81
E)85

(Solution):

$x^3 + 8 = 0 \Rightarrow x^3 = -8 \Rightarrow x = -2$

P(0)=$(-2)^6 - 2(-2)^3 + 1$

=64+16+1

=81

(Answer D)

SUM &SUBSTRACTION ON POLYNOMIALS

(Example):

$P(x)=4x^3 + 6x - 1$

$Q(x) = 6x^3 + 2x + 9$

$\Rightarrow P(x) + Q(x) = ?$

(Solution):

$P(x)+Q(x)=4x^3 + 6x - 1 + 6x^3 + 2x + 9$

$=10x^3 + 5x + 8$

(Example):

$P(x)=4x^4 - 5x^3 - 7$

$Q(x)=-3x^4 + 6x^3 + 3$

$\Rightarrow P(x) - Q(x) = ?$

(Solution):

$P(x)-Q(x)= (4x^4 - 5x^3 - 7) -(3x^4 + 6x^3 + 3)$

$=4x^4 - 5x^3 - 7 + 3x^4 - 6x^3 - 3$

$=7x^4 - 11x^3 - 10$

(Example):

P(x)=$2x^3 - 4x^2$+5x-1

Q(x)=$6x^2 - 4x + 3$

1.P(x)+Q(x)=?

2.P(x)-Q(x)=?

(Solution):

1.P(x)+Q(x)=$\underset{P(x)}{\underline{2x^3-4x^2+5x-1}} + \underset{Q(x)}{\underline{6x^2-4x+3}}$

=$2x^3 + 2x^2 + x + 2$

2.P(x)-Q(x)=$\underset{P(x)}{\underline{2x^3-4x^2+5x-1}} - \underset{Q(x)}{\underline{(6x^2-4x+3)}}$

=$2x^3 - 4x^2 + 5x - 1 - 6x^2 + 4x - 3$

=$2x^3 - 10x^2 + 9x - 4$

(Example):

P(x)=$x^4 - x^3 + 2x^2 + 3x + 2$

Q(x)=$x^4 + 3x^2 - x + 5$

1.P(x)+Q(x)=?

2.P(x)-Q(x)=?

(Solution):

1.P(x)+Q(x)=$\dfrac{\overbrace{x^4-x^3+3x+2}^{P(x)}}{} - \dfrac{\overbrace{x^4+3x^2-x+5}^{Q(x)}}{}$

=$-x^3 + 5x^2 + 2x + 7$

2.P(x)-Q(x)=$\dfrac{\overbrace{x^4-x^3+2x^2+3x+2}^{P(x)}}{} - \dfrac{\overbrace{(-x^4+3x^2-x+5)}^{Q(x)}}{}$

=$x^4 - x^3 + 2x^2 + 3x + 2 + x^4 - 3x^2 + x - 5$

MULTIPLICATION ON POLYNOMIALS

(Example):

$P(x)=x^2 - 3x + 4$

$Q(x)=x^3 - 2x^2 - 1$

$\Rightarrow P(x).Q(x) =?$

(Solution):

$P(x).Q(x)=(x^2 - 3x + 4).(x^3 - 2x^2 - 1)$

$=x^5 - 2x^4 - x^2 - 3x^4 + 6x^3 + 3x + 4x^3 - 8x^2 - 4$

$=x^5 - 5x^4 + 10x^3 - 9x^2 + 3x - 4$

(Example):

$P(x)=x^3 - x^2, Q(x) = x^2 + ax - 9$

$P(x).Q(x)=x^5 + 4x^4 - 14x^3 + 9x^2 \Rightarrow a =?$

$P(x).Q(x)=(x^3 - x^2). (x^2 + ax - 9)$

$=x^5 + ax^4 - 9x^3 - x^4 - ax^3 + 9x^2$

$=x^5 + (a - 1)x^4 - (9 + a)x^3 + 9x^2$

a-1=4$\Rightarrow a = 5$

(Example):

P(x).Q(x)=$x^4 + x^3 - 3x^2 - 4x - 4$

P(x)=$x^2 - 4, Q(x) = x^2 + ax + 1 \Rightarrow a =?$

(Solution):

P(x).Q(x)=$(x^2 - 4). (x^2 + ax + 1)$

=$x^4 + ax^3 + x^2 - 4x^2 - 4ax - 4$

=$x^4 + ax^3 - 3x^2 - 4ax - 4$

=$ax^4 + ax^3 - 3x^2 - 4ax - 4$

=$ax^3 = x^3$

a=1

(Example):

P(x)=$2x^3 - 4x^2 - 3x + 5$

Q(x)=$-3x^4 - 2$

$\Rightarrow P(x). Q(x) =?$

(Solution):

$P(x).Q(x)= 2(x^3 - 4x^2 - 3x + 5).(-3x^4 - 2)$

$=-6x^7 - 4x^3 + 12x^6 + 8x^2+9x^5 + 6x - 15x^4 - 10$

$=-6x^7 + 12x^6 + 9x^5 - 15x^4 - 4x^3 + 8x^2 + 6x - 10$

DIVISION ON POLYNOMIALS

1.(Identity of Division):

$$\frac{P(x)}{Q(x)} \div \frac{Q(x)}{T(x)}$$

P(x)=Q(x).T(x)+K(x)

K(x) is the remainder

(Example):

$$\frac{P(x)}{4} \div \frac{x-3}{Q(x)}, \frac{Q(x)}{2} \div \frac{x+3}{T(x)} \Rightarrow \frac{P(x)}{?} \div x^2 - 9$$

A)3x+4   B)2x-2   c)2x+7

D)4x+1   E)5x+4

(Solution):

$$\frac{P(x)}{4} \div \frac{x-3}{Q(x)}$$

P(x)=(x-3).Q(x)+4

$$\frac{Q(x)}{2} \div \frac{x+3}{T(x)}$$

Q(x)=(x+3).T(x)+2

P(x)=(x-3)[$(x + 3). T(x) + 2$] + 4

$=(x^2 - 9). T(x) + 2x - 6 + 4$

$=(x^2 - 9).$T(x)+2x-2

2x-2(Remainder)

Answer

b

RULE 1:

To find the remainder of P(x) divided by (x+a)is equal

to p(-a).Since (x+a) is a first order polynomial, the remainder must always be equal to real number.

P(x)=(x+a).T(x)+k

P(-a)=(-a+a).T(x)+K=K

$\Rightarrow P(-a) = K$

(Example):

P(x)=$x^3 - 2x^2 + ax + 8$

$\frac{P(x)}{18} \div x - 2 \Rightarrow a =?$

A)1 B)2 C)3

D)4 E)5

221

(Solution):

x-2=0$\Rightarrow x = 2$

P(2)$2^3 - 2.2^2 + a.2 + 8$

18=8-8+2a+8

2a=10$\Rightarrow a = 5$

Answer B

(Example):

P(3x+4)=$x^3 + x^2 - x + 9 \Rightarrow$

$\frac{P(x)}{?} \div x + 2$

A)13   B)9   C)11

D)10   E)7

(Solution):

1.x+2=0$\Rightarrow x = -2$

2.3x+4=-2

3x=-6$\Rightarrow x = -2$

$P(3(-2)+4)=(-2)^3 + (-2)^2 — 2 + 9$

$P(-2)=-8+4+2+9$

$=7$

Answer C

(Example):

$P(4x-1)=x^3 - x^2 + 2x - 5 \Rightarrow$

$\dfrac{P(x)}{?} \div$ x-3

A)2   B)1   C)0

D)-1   E)-3

(Solution):

1.x-3=o$\Rightarrow x = 3$

2.4x-1=3

4x=4$\Rightarrow x = 1$

$P(4.1-1)=1^3 - 1^2 + 2.1 - 5$

$P(3)=1-1+2-5$

=-3

Answer E

(RULE2):(x-a).P(x)=Q(x)$\Rightarrow Q(a) = 0$

(Example):

(x-1)P(x)=$x^4 + ax^3 + 3x - 7 \Rightarrow a =$?

A)2   B)3   C)5

D)7   E)8

(Solution):

x-1=0$\Rightarrow x = 1$

$1^4 + a.1^3 + 3.1 - 7 = (1 - 1)P(1)$

1+a+3-7=0

a=3

Answer B

(Example):

(x-2)P(x)=$x^4 - ax^2 + 2x + 8 \Rightarrow$

$$\frac{\overline{P(x)}}{?} \div x - 1$$

A)-4   B)-3   C)-1

D)3   E)5

224

(Solution):

1.x-2=0$\Rightarrow x = 2$

(2-2)P(2)=16-4a+4+8

0=28-4a

a=7

2.x-1=0$\Rightarrow x = 1$

(1-2)P(1)=1-7+2+8

-1P(1)=4

P(1)=-4

(Remainder)=P(1)=-4

Answer A

(Example):

$$9x^2 + 3x + 7 = (3x + 1).Q(x) + a \Rightarrow a =?$$

A)1   B)2   C)4

D)7   E)9

(Solution):

3x+1=0$\Rightarrow x = -\dfrac{1}{3}$

$$9\left(-\frac{1}{3}\right)^2 + 3\left(-\frac{1}{3}\right) + 7 = \left(3\left(-\frac{1}{3}\right) + 1\right) \cdot Q\left(-\frac{1}{3}\right) + a$$

$$9 \cdot \frac{1}{9} - 3 \cdot \frac{1}{3} + 7 = a$$

a=1-1+7

a=7

RULE 3

To Find the remainder of P(x) divided by $(x^n \pm a)$.

Insert$(x^n = \pm a)$ in the $polynomial$ $P(x)$.

(Example):

$P(x) = x^{16} - 2x^{11} + 6x^6 + 3 \Rightarrow$

$\dfrac{P(x)}{?} \div x^5 + 2$

A)-28x+3   B)4x+9   C)-17x+21

D)-14x+7   E)21x+18

(Solution):

$x^5 + 2 = 0 \Rightarrow x^5 = -2$

$P(x) = x . x^{15} - 2x . x^{10} + 6x . x^5 + 3$

$= x(x^{5^2}) - 2x(x^{5^2}) + 6x(x^6) + 3$

(Remainder)$= x(-2)^3 - 2x(-2)^2 + 6x(-2) + 3$

$= -8x - 8x - 12x + 3$

$= -28x+3$

(Example):

$$(x^2 + 4)P(x) + 8x = ax^2 + 2ax + b + 3 \Rightarrow b = ?$$

A)6   B)8   C)10

D)13   E)15

(Solution):

$$x^2 + 4 = 0 \Rightarrow x^2 = -4$$

((-4)+4)P(x)+8x=a(-4)+2ax+b+3

8x=2ax+b-4a+3

2a=8$\Rightarrow a = 4$

b-4a+3=0

b-16+3=0

b=13

Anwer D

(Example):

P(-1)=10

P(1)=4⇒

$$\frac{\frac{P(x)}{?}}{\div} \div \frac{x^2+1}{x+2}$$

A)2x+1   B)-5x+3   C)-3x+7

D)8x+1   E)-2x-4

(Solution):

P(x)=$(x^2 + 1)(x + 2) + ax + b$

X=1⇒ $P(1)6 + a + b = 4$ ⇒ $a + b = -2$

X=-1⇒ $P(-1) = 2 - a + b = 10 \Rightarrow \frac{b-a=8}{\begin{array}{c}2b=6\\b=3\end{array}}$

a=5⇒ $ax + b = -5x + 3$

Answer B

(Example):

P(x)=$x^3 - 3x^2 + 4x - 9$

$$\frac{P(x)}{?} \div x^2 - x + 1$$

A)2x+5   B)x+9   C)x-7

D)3x-5   E)5x+8

(Solution):

$$x^2 - x + 1 = 0 \Rightarrow x^2 = x - 1$$

(if we write x-1 instead of x^2 in $P(x)$.)

P(x)=x.$x^2 - 3x^2 + 4x - 9$

(Remainder)=x(x-1)-3(x-1)+4x-9

$$=x^2 - x - 3x + 3 + 4x - 9$$

=x-1-6=x-7

Answer C

RULE 4

1.The remainder of P(x) and Q(x) divided by (x-a) are

Equal to A and B, respectively.

a)The remainder P(x)$\pm$Q(x) devided by (x-a) is equal to

A$\pm$B.

b)The remainder P(x).Q(x) divided by (x-a) equal to

A.B

2.The remainder of P(x) divided by (x-a) and (x-b) are

Equal to A and B, respectively. Then, the remainder

P(x) divided by (x-a).(x-b) is in the form of to mx+n).

(Example)

$$\frac{\overline{P(x)}}{3} \div x - 4$$

$$\frac{\overline{Q(x)}}{4} \div x - 4$$

$$\frac{x.P(x)+(x+1).Q(x).x^2+2x}{?} \div x - 4$$

A)56   B)48   C)46

D)44   E)40

(Solution):

1.x-4=0$\longrightarrow x = 4$

P(4)=3.Q(4)=4

2.(Remainder)=4P(4)+(4+1).Q(4)+$4^2 + 2.4$

=4.3+5.4+16+8

12+20+24

=56

                       Answer A

(Example):

$$\frac{\overline{P(x)}}{12x+7} \div (3x - 4)^2 \Rightarrow \frac{\overline{P(x)}}{?} \div 3x - 4$$

A)15   B)17   C)19

D)21   E)23

(Solution):

$3x-4=0 \Rightarrow x = \frac{4}{3}$

$P(x)=(3x - 4)^2 Q(x) + 12x + 7$

$X=\frac{4}{3}$

$P(\frac{4}{3}) = \frac{(\frac{4}{3}.3-4)^2}{0} Q\left(\frac{4}{3}\right) + 12\frac{4}{3} + 7$

(Remainder)=0+4.4+7

$=23$

Answer E

(Example):

$$\frac{P(x)}{5} \div x - 2$$

$$\frac{Q(x)}{9} \div x - 3 \Rightarrow \frac{P(x)}{?} \div (x - 2)(x - 3)$$

A)2x-7   B)4x+8   C)4x-3

D)8x+9   E)12x-1

(Solution):

K(x)=mx+n

$x-2=0 \Rightarrow x = 2 \Rightarrow 2m + n = 5$

$x-3=0 \Rightarrow x = 3 \Rightarrow \frac{-3m+n=9}{-m=-4}$

m=4(and)n=-3

K(x)=4x-3

Answer C

(Example):

P(x)=$x^3 - x^2 + ax + b$

$\dfrac{P(x)}{0} \div x^2 - 3x + 2 \Rightarrow a + b =?$

A)5 B)4 C)2

D)0 E)-2

(Solution):

$x^2 - 3x + 2 = (x - 1)(x - 2)$

1.X-1=0$\Rightarrow x = 1$

P(1)=1-1+a+b=0$\Rightarrow a = -b$

2.x-2=0$\Rightarrow x = 2$

P(2)=8-4+2a+b=0

2a+b+4=0

-2b+b+4=0

b=4,a=-4

a+b=4(-4)=0

Answer D

TEST WITH SOLUTION

1. $P(x)=x+4, Q(x)=x^2 - 5x \Rightarrow P(x) + Q(x) = ?$

A) $x^2 - 4x$ B) $x^2 - 4$ C) $x^2 + 4x + 4$

D) $(x - 2)^2$ E) $(x + 4)^2$

(Solution):

$P(x)+Q(x)=\underset{P(x)}{\underbrace{x+4}} + \underset{Q(x)}{\underbrace{x^2-5x}}$

$=x^2 - 4x + 4$

$=(x - 2)^2$

Answer D

2. $(x^3 - 4x^2 + 3x).(x^2 - 5x + 1) = \cdots + a.x^4 + \cdots$

$\Rightarrow a = ?$

A)-11   B)-10   C)-9

D)-8   E)-7

(Solution):

$(x^3 - 4x^2 + 3x).(x^2 - 5x + 1)$

$-5x^4 - 4x^4 = ax^4$

$-9x^4 = ax^4$

a=-9

3.P(x-4)=$2x^2 + 3x + 4 \Rightarrow P(1) = ?$

A)69   B)70   C)71

D)72   E)73

(Solution):

P(x-4)=$2x^2 + 3x + 4$

$\Rightarrow x - 4 = 1 \Rightarrow x = 5$

P(5-4)=$2.5^2 + 3.5 + 4$

=2.25+15+4

=50+15+4

=69

Answer A

4.P(x-3)=$x^3 + 2x^2 - x + a, \quad P(-1) = 5 \Rightarrow a = ?$

A)-10   B)-9   C)-8

D)-7   E)-6

(Solution):

$P(x-3)=x^3 + 2x^2 - x + a$

$X=2 \Rightarrow$

$P(2-3)=2^3 + 2.2^2 - 2 + a$

P(-1)=8+2.4-2+a

P(-1)=8+8-2+a

P(-1)=14+a

$5=14+a \Rightarrow a = -9$

Answer B

$5.P(x^2) = x^4 - 1. Q(\sqrt{x}) = x + 1 \Rightarrow P(x).Q(x) =?$

A)$x^4 + 1$ B) $x^4 - 1$ C) $x^2 - 1$

D) $x^2 + 1$ E) $x^5 - 1$

(Solution):

$P(x^2) = x^{2^2} - 1$ $Q(\sqrt{x^2}) = x^2 + 1$

P(x)=$x^2 - 1$ $Q(x) = x^2 + 1$

P(x).Q(x)=$(x^2 - 1).(x^2 + 1)$

$=x^4 + x^2 - x^2 - 1$

$=x^4 - 1$

Answer B

237

6.m∈ Z

$P(x)=x^{2m+2} + x^{2m+1} + 2 \Rightarrow P(-1) =?$

A)2   B)3   C)4

D)5   E)6

(Solution):

$P(x)=x^{2m+2} + x^{2m+1} + 2$

$P(-1)=(-1)^{2m+2} + (-1)^{2m+1} + 2$

=1+(-1)+2

=2

Answer A

7.$P(2-3x)=-2x^7 + 5x^3 + 2x^2 + 8 \Rightarrow P(5) =?$

A)3   B)4   C)5

D)6   E)7

(Solution):

$P(2-3x)=-2x^7 + 5x^3 + 2x^2 + 8$

X=-1$\Rightarrow$

$P(2-3(-1))=-2.(-1)^7 + 5.(-1)^3 + 2.(-1)^2 + 8$

P(2+3)=-2.(-1)+5.(-1)+2.1+8

P(5)=2-5+2+8

  =7

Answer E

8.P(x)=$2x^4 - ax^3 + x^2 - (3 + b)x + 1$

Q(x)=$(c+1)x^4 - 2x^2 + 2x + 3,$

P(x)+Q(x)=$-x^2 + 4 \Rightarrow$ $\qquad a + b + c =?$

A)-7   B)-6   C)-5

D)-4   E)-3

(Solution):

P(x)+Q(x)=$(c+1+2)x^4 - ax^3 - x^2 + (2 - 3 - b)x + 4$

=(c+3) $x^4 - ax^3 - x^2 + (-1 - b)x + 4$

(c+3) $x^4 - ax^3 - x^2 + (-1 - b)x + 4 = x^2 + 4$

C+3=0   -a=0   -1-b=0

c=-3   a=0   b=-1

a+b+c=0+(-1)+(-3)

=-4

9.P(x,y)=$2x^2y^2 - 3xy^2 - 6x + 1 \Rightarrow p(3, \sqrt{2}) =$?

A)1   B)2   C)3

D)4   E)5

(Solution):

$P(3.\sqrt{2}) = 2.3^2.(\sqrt{2})^2 - 3.3(\sqrt{2})^2 - 6.3 + 1$

=2.9.2-9.2-18+1

=36-18-18+1

=1

Answer A

10.P(x)=$x^3 + ax^2 - bx + 1, P(1) = 10 \Rightarrow a - b =$?

A)4   B)5   C)6

D)7   E)8

(Solution):

P(1)=$1^3 + a.1^2 - b.1 + 1 = 10$

1+a-b+1=10

a-b=8

Answer E

11. $P(x)=x^4 - 5x^2 + 4 \Rightarrow \frac{P(x)-Q(x)}{(x^2+1).(x+1)}$

$Q(x)=x^3 - 4x^2 + x + 6$

A)x-1   B)x+1   C)x-2

D)x+3   E)x-3

(Solution):

$\frac{P(x)-Q(x)}{(x^2+1).(x+1)}$

$=\frac{x^4-5x^2+4-(x^3-4x^2+x+6)}{(x^2+1).(x+1)}$

$=\frac{x^4-5x^2+4-x^3+4x^2-x-6}{(x^2+1).(x+1)}$

$=\frac{x^4-x^2-2-x^3-x}{(x^2+1).(x+1)}$

$=\frac{(x^2-2).(x^2+1)-x(x^2+1)}{(x^2+1).(x+1)}$

$=\frac{x^2-x-2}{x+1}$

$=\frac{(x-2)(x+1)}{x+2}$

$=x-2$

Answer E

12. $P(x)=(2k-1)x^3 + (k + 1)x^2 - x + k$

P(-2)=36$\Rightarrow k =$?

A)-3   B)-2   C)0

D)2   E)3

(Solution):

P(-2)=(2k-1).$(-2)^3 + (k + 1)(-2)^2 - 2 + k$

=(2k-1).(-8)+(k+1).4+2+k

=-16k+8+4k+4+2+k

=-11k+14

-11k+14=36

-11k=22

K=-2

$\qquad\qquad$ Answer B

13.$(x^2 + x + 1).P(x + 5) = x^3 - 1 \Rightarrow P(x) =$?

A)x+6   B)x-5   C)x-6

D)x+5   E)x-1

(Solution):

$(x^2 + x + 1).P(x + 5) = x^3 - 1$

$$P(x+5)=\frac{x^3-1}{x^2+x+1}$$

$$P(x+5)=\frac{(x-1).(x^2+x+1)}{x^2+x+1}$$

P(x+5)=x-1

P(x-5+5)=x-5-1

P(x)=x-6

Answer C

14.m,n∈ Z^+

$$4x^2 - mx + 4 = (2x - n)^2 \Rightarrow m + n =?$$

A)10   B)12   C)14

D)16   E)18

(Solution):

$$4x^2 - mx + 4 = (2x - n)^2$$

$$4x^2 - mx + 4 = 4x^2 - 4nx + n^2$$

-mx+4=-4nx+n^2

-m=-4n (and)$n^2 = 4, n = 2$

n=2⇒ $m = 8 \Rightarrow m + n = 10$

15.$x^3 + ax^2 + bx + c = (x - 2).(x + 4).(x + 1)$

$\Rightarrow a.b.c =?$

A)96   B)108   C)120

D)132   E)144

(Solution):

$x^3 + ax^2 + bx + c = (x - 2).(x + 4).(x + 1)$

$$=(x^2 + 4x - 2x - 8).(x + 1)$$

$$=(x^2 + 2x - 8)(x + 1)$$

$$=x^3 + x^2 + 2x^2 + 2x - 8x - 8$$

$$=x^3 + 3x^2 - 6x - 8$$

$x^3 + ax^2 + bx + c = x^3 + 3x^2 - 6x - 8$

a=3, b=-6,c=-8

a.b.c=3.(-6).(-8)

=144

Answer E

16.P(x)=-3.x^{40} + 12.x^{20} - 12 $\Rightarrow P(\sqrt[5]{2}) =?$

A)-768　　　　　　　B)-640　　　　　　　C)-612

D)-588　　　　　　　E)-542

(Solution):

$P(x)=3.x^{5^6} + 12.(x^{5^4}) - 12$

$X=\sqrt[5]{2} \Rightarrow$

$P(\sqrt[5]{2})=-3((\sqrt[5]{2}^5)^8 + 12((\sqrt[5]{2}^5)^2 - 12$

$=-3.2^8 + 12.2^4 - 12$

=-588

Answer D

17.$P(2x+5)=(2x^2 + 3x - 1).Q(x + 1)$

$Q(-1)=3 \Rightarrow P(1) =?$

A)1　　　　　　　B)2　　　　　　　C)3

D)4　　　　　　　E)5

(Solution):

P(2x+5)=(2x+3x-1).Q(x+1)

X=-2 (for)

$P(2.(-2)+5)=(2.(-2)^2 + 3.(-2) - 1).Q(-2 + 1)$

P(1)=(2.4-6-1).Q(-1)

P(1)=Q(-1)

P(1)=3

Answer C

18.P(x)=$ax^2 + bx + c, P(2) = 0, P(3) = 0$

$\Rightarrow \frac{a}{b} =?$

A)$-\frac{1}{10}$ B) $-\frac{1}{5}$ C)$\frac{1}{5}$

D)$\frac{1}{10}$ E)$\frac{1}{2}$

(Solution):

P(x)=$ax^2 + bx + c$, $P(2) = a.2^2 + b.2 + c$

=4a+2b+c$\Rightarrow 4a + 2b + c = 0$

P(3)=a.$3^2 + b.3 + c$

=9a+3b+c$\Rightarrow 9a + 3b + c = 0$

$\qquad 4a + 3b + c = 0$

$\qquad ....$

$\qquad -9a - 3b - c = 0$

$\qquad 4a + 2b + c = 0$

$\qquad ....$

$$-5a - b = 0$$

-5a=b

$$\frac{a}{b} = -\frac{1}{5}$$

Answer B

19.P(x)=$-x^2 + 3x.$ $Q(x) = 5x^2 - x + 2$

$\Rightarrow P[Q(1)] + Q[P(-1)] =?$

A)36   B)48   C)50

D)56   E)68

(Solution):

Q(1)=5.1-1+2   ,P(-1)=-$(-1)^2 + 3.(-1)$

Q(1)=6   ,P(-1)=-1-3

                         P(-1)=-4

P[Q(1)] + Q[P(-1)] = P(6) + Q(-4)

=$-6^2 + 3.6 + 5.(-4)^2 - 4 + 2$

=-36+18+5.16+4+2

=-36+18+80+4+2

=68

20.$P(x+3)=2x^3 + ax^2 + x + 1, P(4) = 10 \Rightarrow a =?$

A)-6   B)-2   C)2

D)6   E)10

(Solution):

$P(x+3)=2x^3 + ax^2 + x + 1, x = 1 (for)$

$P(1+3)=2.1^3 + a.1^2 + 1 + 1$

P(4)=2+a+2

P(4)=4+a

4+a=10

a=6

Answer D

21.$P(x^2 + 2) = x^8 + x^6 + ax^4 + 3$

$P(0)=47 \Rightarrow a =?$

A)7   B)8   C)9

D)10   E)11

(Solution):

P(0)=47

$$x^2 + 2 = 0 \Rightarrow x^2 = -2$$

P(-2+2)=$(-2)^4 + (-2)^3 + a(-2)^2 + 3$

47=16-8+4a+3

47=11+4a

36=4a

a=9

Answer C

22.P(x+1)-5=xP(x)+Q(x)

P(1)=35

$$\Rightarrow Q(0) =?$$

A)30   B)37   C)42

D)47   E)55

(Solution):

P(1)=35

Q(0)=?$\Rightarrow$

P(1)-5=0.P(0)+Q(0)

35-5=Q(0)

Q(0)=30

Answer A

QUESTIONS

1.P(x)=$x^3 + 3x^2 + x, Q(x) = 5x^2 + bx + 1$

P(x)+Q(x)=$x^3 + 8x^2 + 5x + 1 \Rightarrow b =?$

A)1   B)2   C)3

D)4   E)5

(Solution):

P(x)+Q(x)=$x^3 + 3x^2 + x + 5x^2 + bx + 1$

=$x^3 + 8x^2 + bx + x + 1$

= $x^3 + 8x^2 + (b + 1)x + 1$

$x^3 + 8x^2 + 5x + 1 = x^3 + 8x^2 + (b + 1)x + 1$

5=b+1$\Rightarrow b = 4$

Answer D

2.P(x)=x+1,Q(x)=$x^2 + x$

$\Rightarrow P(x) + Q(x) =?$

A)$x^2 - 2x$ B) $x^2 + 2x$ C)(x-1)(x+1)

D)$(x + 1)^2$ E)$(x - 1)^2$

(Solution):

P(x)+Q(x)=x+1+$x^2 + x$

$\qquad =x^2 + 2x + 1$

$\qquad =(x + 1)^2$

Answer D

3.P(x)=$x^2 + 5x - 3$, $Q(x) = x + 1$

$\Rightarrow P[Q(-1)] + Q[P(2)] =?$

A)14   B)11   C)9

D)-6   E)-4

(Solution):

Q(-1)=-1+1=0,P(2)=$2^2 + 5.2 - 3$

P(2)=4+10-3

P(2)=11

$$P[Q(-1)] + Q[P(2)] = P(0) + Q(11)$$

$$=0^2 + 5.0 - 3 + 11 + 1$$

=-3+12=9

Answer C

4.P(x)=$ax^2 + bx + c$,$P(1) = 0, P(2) = 0$

$\Rightarrow \frac{b}{a} =?$

A)3   B)2   C)0

D)-2   E)-3

(Solution):

$P(1)=a.1^2 + b.1 + c = 0 \Rightarrow -1/a + b + c = 0$

$P(2)=a2^2 + b.2 + c = 0 \Rightarrow$ $\begin{array}{r} 4a+2b+c=0 \\ -a-b-c=0 \\ +4a+2b+c=0 \\ \hline 3a+b=0 \end{array}$

b=-3a

$\frac{b}{a} = -3$

Answer E

5.$x^3 + ax^2 + bx + c = (x + 2).(x - 3).(x - 1)$

252

$$\Rightarrow \frac{c}{a} = ?$$

A)3   B)2   C)1

D)-1   E)-3

(Solution):

$$x^3 + ax^2 + bx + c = (x + 2)(x - 3)(x - 1)$$

$$= (x^2 - 3x + 2x - 6)(x - 1)$$

$$= x^3 - x^2 - x^2 + x - 6x + 6$$

$$x^3 + ax^2 + bx + c = x^3 - 2x^2 - 5x + 6$$

a=-2,b=-5,c=6, $\frac{c}{a} = \frac{6}{-2} = -3$

Answer E

6.a,b$\in R$ $4x^2 - 5x + b = (2x - a)^2 = 0$

a+b=?

A)$\frac{5}{4}$ B)$\frac{25}{8}$ C)$\frac{25}{16}$

D)$\frac{35}{8}$ E)$\frac{45}{16}$

(Solution):

$$4x^2 - 5x + b = (2x - a)^2$$

$$4x^2 - 5x + b = 4x^2 - 4ax + a^2$$

-5=4a

$$a=\frac{5}{4}$$

$$b=a^2$$

$$b=\left(\frac{5}{4}\right)^2 \Rightarrow b = \frac{25}{16}$$

$$\text{a+b}=\frac{5}{4} + \frac{25}{16} = \frac{20}{16} + \frac{25}{16} = \frac{45}{16}$$

(4)

Answer E

$$7.\frac{3x^2-2mx^2-nx-2}{0} \div \frac{x^2-x-2}{Q(x)} \Rightarrow m =?$$

(Solution):

$$3x^2 - 2mx^2 - nx - 2 = (x - 2)(x + 1). Q(x)$$

$$\text{X=2} \Rightarrow 24 - 8m - 2n - 2 = 0$$

$$\Rightarrow 11 = 4m + n$$

$$\text{X=-1} \Rightarrow -3 - 2m + n - 2 = 0$$

$$\Rightarrow 2m - n = -5$$

4m+n=11

2m-n=-5

....

6m=6

m=1

Answer D

1. $(2x + 1)^3 = 8x^3 + ax^2 + bx + c \Rightarrow a + b + c = ?$

A)12   B)14   C)16

D)19   E)21

2. $(a+x).(x^2 + ax + 2b) = x^3 + 3x^2 + 5x + c + 2$

$\Rightarrow 2a + c = ?$

A)5   B)6   C)7

D)8   E)9

3. P(x)=$3x^5 - x - 1 \Rightarrow P(1) = ?$

A)0   B)1   C)2

D)3   E)4

4. P(1-x)=$2x^2 + 2 \Rightarrow P(X) = ?$

A)$2x^2 - 4$ B) $2x^2 + 4x$ C) $2x^2 - 4x$

D) $2x^2 + 4x + 4$ E) $2x^2 - 4x + 4$

5. P(x-2)=ax+b-2a$\Rightarrow P(2 - x) = ?$

A)-ax-b   B)ax-b   C)-ax+b+2a

256

D)-ax+2b+a   E)ax-b+2a

6.P(x-2)=$ax^6 - bx^3 + cx^2 - dx - 9$

P(-3)=6$\Rightarrow a + b + c + d =$?

A)15   B)16   C)17   D)18   E)19

7.$P(x^3 + 1) = x^8 + +a.x^6 - 3x^4 + 2$

P(0)=20$\Rightarrow a =$?

A)20   B)15   C)10

D)5   E)2

8.P(x)=$x^7 + 4x + a, P(2) = 0 \Rightarrow$?

A)-136   B)-120   C)-90

D)136   E)140

9.P(x+2)=$2x^4 - 3x - 2$

Q(x)=$3x^2 - 2x + 2 \Rightarrow P(0).Q(0) =$?

A)56   B)64   C)72

D)76   E)80

10. P(x)=(3x-4).Q(x)+3,   P(4)=19 $\Rightarrow Q(4) =$?

A)1   B)2   C)3

D)4   E)5

11. P(x-2)=(x-2).Q(x+2)+x+3, P(2)=15

$\Rightarrow Q(6) =$?

A)1   B)2   C)3

D)4   E)5

12. P(x)=ax+b $\Rightarrow P(1) - P(2) =$?

A)-a   B)-b   C)2b

D)a   E)b

13. (x+3).P(x+3)+2=$x^3 ax + 5 \Rightarrow P(3) =$?

A)-2   B)-1   C)0   D)1   E)2

14. P(x+1)=$(x^2 + 2x - 1). Q(x) + x - 2$

P(2)=11 $\Rightarrow Q(1) =$?

A)2   B)3   C)4

D)5   E)6

15. $P(x+3)=x^3 - 2x - 3 \Rightarrow P(-2) =?$

A)-156   B)-144   C)-118

D)118   E)144

16. $P(x.y)=x^3.y - x^2.y^2 + 2.y^4$

$\Rightarrow P(\sqrt{5}.-\sqrt{5}) =?$

A)-25   B)-5   C)0

D)-2   E)25

17. $P(x^2) = x^4 + 5x^2 + 8 \Rightarrow P(-3) =?$

A)1   B)2   C)3

D)-2   E)-3

18. $P(2x+1)=2x+5 \Rightarrow P(x) =?$

A)x+2   B)x+4   C)x+5

D)2x+1   E)x.(x+3)

19. $P(x)=2x^2 - 2x + 1$

$P(x+2)=P(x-2) \Rightarrow x =?$

A)$-\frac{1}{4}$

B) $-\frac{1}{8}$

C)$\frac{1}{8}$

D)$\frac{1}{4}$

E)$\frac{1}{2}$

20.P(x)=$x^3 + 6x^2 + 12x + 8 \qquad \Rightarrow P(x - 2) =?$

A)$x^3 + 2$

B) $x^3 + 8$

C) x^3

D) $x^3 - 2$

E)$x^3 - 8$

21.$\frac{P(x+2)}{Q(X)} = 3x^2 - x - 15, \qquad Q(-3) = 4$

$\Rightarrow P(-1) =?$

A)30

B)40

C)50

D)2

22.P(x-1)=$2x^2 + ax + b$

P(x+1)=$2x^2 + x + 1 \Rightarrow a.b =?$

A)-49

B)-21

C)-15

D)2

E)1

23.P(x)=$(x - 2)^3 \Rightarrow P\left(\sqrt[3]{3} + 2\right) =?$

A)-3

B)-4

C)3

D)2

E)1

24.P(x+1)=$x^2 - 4x + 7 \qquad \Rightarrow P(1) =?$

A)9   B)8   C)7   D)6   E)5

25. $P(x)=x^3 - 3x^2 + 3x - 1 \Rightarrow P(x+1) =?$

A)x^3 B)$2x-x^3$ C)$x^3 - 1$ D)$x^3 + 2x$ E)$1 - x^3$

(Answers)

1.D	2.B	3.B	4.E	5.C	6.A
7.A	8.A	9.C	10.B	11.D	12.A
13.D	14.E	15.C	16.A	17.B	18.B
19.E	20.C	21.D	22.A	23.C	24.C
25.A					

1. $P(x)=x^2 + 2x$

$Q(x)$=x-3 $\Rightarrow P[Q(4)] + 2.Q[P(1)] =?$

A)0   B)1   C)2

D)3   4)4

2. $P(2x+1)$=4x+5 $\Rightarrow P(3) =?$

A)-1   B)4   C)5

D)9   E)11

3. $P(x+2)=4x^2 + mx + 5$

P(4)=51 $\Rightarrow m =?$

A)2   B)4   C)5

D)6   E)7

4. $P(x)=2x^3 + (m - 2)x^2 + 4x + 5$

P(x)=(x-3).Q(x)+116

A)4   B)5   C)6

D)7   E)8

5. $P(x+1)=x^3 - 4x^2 + x + m$

262

P(x+1)=(x-2).Q(x)+10⇒ $m =?$

A)6 B)10 C)12

D)16 E)18

6.P(x+1)=$x^2 - 3x \Rightarrow P(x) =?$

A)$x^2 - 2x + 1$ B)$x^2 - 5x + 4$
C)$2x^2 - 6x + 3$

D)$x^2 - 4x + 2$ E)$x^2 - 5x - 2$

7.P(2x+4)=$x^3 - 4x^2 + 1 \Rightarrow P(5) =?$

A)1 B)$\frac{1}{2}$ C)$\frac{1}{4}$

D)$\frac{1}{4}$ E)$\frac{1}{64}$

8.P(x)=$2\sqrt{2}x + 12 \Rightarrow P(\sqrt{2}) =?$

A)5 B)8 C)16

D)$4\sqrt{2}$ E)$\sqrt{2} + 12$

9.$ax^3 + 2x^2 + bx + c = (x + 1).(x - 2).(x + 3)$

$\Rightarrow a + b + c =?$

A)-9 B)-10 C)13

D)14 E)21

263

10. $P(x,y)=4x^2y^3 - 2x^2 + y^2x + 16 \Rightarrow P(-1,1) =?$

A)-2   B)7   C)17

D)19   E)20

11. $P(x)=x^3 - ax^2 + bx + 7$

$P(2)=0$

$P(1)=4 \Rightarrow a =?$

A)$\frac{1}{2}$ B)$\frac{2}{3}$ C)$\frac{7}{2}$

D)4   E)16

12. $P(x)=9x^2 + 8x$

$Q(x)=4x+3 \Rightarrow P[Q(2)] - Q[P(3)] =?$

A)211   B)105   C)754   D)801
E)902

13. $P(x)=6x^2 + 4x + 3 + b$

   $P(x)=(x-1).Q(x)+27 \Rightarrow b =?$

A)12   B)13   C)14

D)15   E)16

14. $P(x)=2x^2 + 4x - 10 \Rightarrow P(-3) =?$

A)-5   B)13   C)14

D)15   E)16

15. $P(x)=4x^2 + 7x - 8$

$Q(x)=3x^3 + 5x^2 - 4x - 7$

$P(x)+Q(x)=T(x) \Rightarrow T(1) =?$

A)-2   B)-1   C)0

D)15   E)24

16. $P(x-2)=4x^3 + 5x^2 - 6 \Rightarrow P(-1) =?$

A)-4   B)3   C)2

D)4   E)12

17. $P(x)=2x^2 + ax + b$

$P(1)=0 \Rightarrow a + b =?$

A)-2   B)4   C)10

D)12   E)14

18. $P(x+2)=3x^3 + ax^2 + x + 1$

P(3)=8$\Rightarrow a =$?

A)0 　　　　 B)1 　　　　　　 C)2 　　　 D)3 　　　　　 E)4

19.P(x)=$x^2 + 4x - 5 \Rightarrow \frac{P(x)+Q(x)}{(x-1)(x+1)} =$?

Q(x)=4-4x

A)0 　　　　 B)1 　　　　　 C)2

D)3 　　　　 E)4

20.P(x)=4x+3

Q(x)=2x+1$\Rightarrow P(x).Q(x) =$?

A)$2x^2 + 5x + 4$ 　　　　　 B) $8x^2 + 10x + 3$
C) $5x^2 - 2x + 3$

D)$-8x^2 - 5x + 3$ 　　　　 E) $2x^2 - 10x + 5$

21.P(x)=$x^2 + 4x \Rightarrow P\big(P(-4)\big) =$?

A)0 　　　　 B)1 　　　　　 C)2 　　　　 D)3
E)4

22.P(x)=$(x^2 - 3x + 2)$, 　　 $Q(x) = 4x^3 + x^2$

$\Rightarrow P(1) + Q(2) =$?

A)32 　　　　 B)36 　　　　 C)47 　　　 D)54
E)60

23. $P(x)=x^3 + 1 \quad \dfrac{-\frac{P(x)}{2} \div \frac{x-1}{Q(X)}}{}$

$\Rightarrow Q(0) =?$

A)$\frac{1}{2}$ B)1   C)-$\frac{1}{2}$ D)-1   E)-2

24. $P(x)=ax^2 + bx + c$

$P(0)=2$ $\Rightarrow a + b + c =?$

$P(1)=8$

A)8   B)6   C)4

D)2   E)0

(Answers)

1.D	2.D	3.E	4.D	5.D	6.E
7.D	8.C	9.A	10.C	11.C	12.C
13.C	14.B	15.C	16.B	17.A	18.D
19.B	20.B	21.A	22.B	23.D	24.C

1.$P(x+1)=x^2 - 4x + 5 \Rightarrow P(2 - x) =$?

A)$x^2 - x + 2$ B) $x^2 + 2x + 2$
C) $x^2 + x - 2$

D) $x^2 - x + 1$ E) $x^2 - 2x - 1$

2.$P(x-2)=x^2 + x - 6 \Rightarrow P(x + 2) =$?

A) $x^2 + 9x + 14$ B) $x^2 + 7x + 7$
C) $x^2 - 9x - 14$

D) $x^2 - 7x + 7$ E) $x^2 + 8x$

3.$P(x,y)=x^3 + 3x^2y + 3xy^2 + y^3$

$\Rightarrow P\left(\sqrt[3]{4} + 2, \sqrt[3]{4} - 2\right) =$?

A)64   B)32   C)16   D)8
E)4

4.$P(x,y)=x^2 - 4xy + 4y^2$

$\Rightarrow P\left(2\sqrt{2}, 2\sqrt{2}\right) =$?

A)12   B)8   C)4   D)2
E)0

5.$P(x+2)=ax^2 + 6x^2 + 3x - 2$

P(1)=2a-5 $\Rightarrow P(3) =?$

A)1   B)-1   C)-2   D)5
E)9

6.P(x+1)=$mx^2 + x + 1$

P(0)=3 $\Rightarrow P(-1) =?$

A)11   B)10   C)7   D)5
E)1

7.P(x+2)=$-3x^4 + 2x^2 + 4x - 2$

    $\Rightarrow P(1) - P(0) =?$

A)43   B)18   C)8   D)4   E)2

8.P(2x-3)=$-8x^3 + 2x + 4$ $\Rightarrow P(0) =?$

A)-28   B)-20   C)-12   D)-4   E)8

9.P$(\frac{x}{2}) = 5x^5 + 4x^3 + 15x - 7$

    $\Rightarrow P(0) =?$

A)-5   B)-6   C)-7   D)-8   E)-9

10. $P(x-2)=\dfrac{x^2-5x-8}{Q(2x-1)} - x$

   $P(-1)=23 \quad \Rightarrow Q(1) =?$

A)18   B)7   C)-3   D)$-\dfrac{1}{2}$ E)$-\dfrac{7}{2}$

11. $P(x-1).Q(x+3)=x^3 - 3x^2 - 4$

   $P(0)=-3 \Rightarrow \quad Q(4) =?$

A)-3   B)-2   C)0   D)1   E)2

12. $P(x-2)=2x^2 - 3x - 6$

   $\Rightarrow P(\sqrt{2}) =?$

A)$-\sqrt{2} + 1$ B)$-3\sqrt{2}$ C)$2\sqrt{2}$ D)$4\sqrt{2} + 1$
E)$5\sqrt{2}$

13. $P(x+1)=x^2 + x + 1$

   $P(\sqrt{3}) =?$

A)$5\sqrt{3}$ B)$3+\sqrt{3}$ C)$2+\sqrt{3}$ D)$\sqrt{3}$
E)$4-\sqrt{3}$

14. $P(x-2)=x^2 - 4x + 4$

$$\Rightarrow P(x+2) =?$$

A)$x^2 + 4x + 4$ B) $x^2 - 3x + 4$

C) $x^2 - 2x - 3$

D) $x^2 + 4$ E) $x^2 - 5x + 3$

15. $\dfrac{25x-9}{x^2+4} = \dfrac{A}{x-1} + \dfrac{B}{x+1}$

$$\Rightarrow A + B =?$$

A)5   B)8   C)12   D)17

E)25

16.P(x)= $x^2 - 3x + 1$ $\Rightarrow P(2x - 4) =?$

A) $3x^2 - 8x - 7$ B) $4x^2 + 5x + 9$

C) $4x^2 - 22x + 29$

D) $5x^2 + 4x + 10$ E) $2x^2 + x + 17$

17.P(3x+8)=$5x^2 + 3x - 4$

    $\Rightarrow P(-1) =?$

A)32   B)24   C)18   D)8

E)2

18.P(1-x)=-3x+5 $\Rightarrow P[P(2)]=?$

A)24   B)26   C)32   D)48
E)57

19.P(2x+2)=$x^2 + 1 \Rightarrow P\big(P(0)\big) =?$

A)-3   B)1   C)5   D)8
E)17

20.$\dfrac{-7x+7}{x^2-5x+6} = \dfrac{A}{x-3} + \dfrac{B}{X-2} \Rightarrow A + 2B =?$

A)-5   B)0   C)7   D)17
E)20

21.$\dfrac{P(x+1)}{Q(2x+1)} = x^2 + 10x + 3$

 P(0)=-18$\Rightarrow Q(-1) =?$

A)-1   B)3   C)4   D)5
E)6

22.P(2x-3)=4x-6$\Rightarrow P(2x) =?$

A)2x   B)4x   C)6x   D)8x
E)9x

23.P(x-2)=Q(x).$(x^2 - x - 2) + x^3$

272

$$\Rightarrow P(0) =?$$

A)2   B)4   C)6   D)8

E)10

24. $\dfrac{\dfrac{P(x)}{13}}{} \div \dfrac{x+2}{x^2-3} \Rightarrow \dfrac{P(0)}{P(-2)} =?$

A)$\dfrac{7}{13}$ B)3   C)$\dfrac{10}{3}$ D)$\dfrac{13}{2}$

E)10

(Answers)

1.B	2.A	3.B	4.B	5.E	6.A
7.A	8.B	9.C	10.D	11.E	12.E
13.E	14.A	15.E	16.C	17.A	18.B
19.B	20.B	21.B	22.B	23.D	24.A

Printed in Great Britain
by Amazon